Decision Engineering

Series Editor

Rajkumar Roy

For further volumes:
http://www.springer.com/series/5112

Andrew C. Lyons
Adrian E. Coronado Mondragon
Frank Piller · Raúl Poler

Customer-Driven Supply Chains

From Glass Pipelines to Open Innovation Networks

Andrew C. Lyons
University of Liverpool
Management School
Chatham Street
Liverpool L69 7ZH
UK
e-mail: A.C.Lyons@liverpool.ac.uk

Adrian E. Coronado Mondragon
Royal Holloway
University of London
London
UK
e-mail: adrian.coronado@rhul.ac.uk

Frank Piller
RWTH Aachen University
Technology and Innovation Management Group
Templergraben 64
52062 Aachen
Germany
e-mail: piller@tim.rwth-aachen.de

Raúl Poler
CIGIP
Universidad Politécnica de Valencia
Camino de Vera s/n
46022 Valencia
Spain
e-mail: rpoler@cigip.upv.es

ISSN 1619-5736
ISBN 978-1-84628-875-3   e-ISBN 978-1-84628-876-0
DOI 10.1007/978-1-84628-876-0
Springer London Dordrecht Heidelberg New York

British Library Cataloguing in Publication Data
A catalogue record for this book is available from the British Library

Library of Congress Control Number: 2011941717

© Springer-Verlag London Limited 2012

© 2005-2011 Pandora Media, Inc., All Rights Reserved Pandora and the Music
Genome Project are registered trademarks of Pandora Media, Inc.
Mini Cooper © Copyright 2008 BMW AG, Munich, Germany
Kenmore © 2011 Sears Brands, LLC. All Rights Reserved.
Adidas® 2011 adidas Group
Copyright © zafu - we match jeans and bras to you 2011. All Rights Reserved.
Fedex © FedEx Office and Print Services, Inc. Three Galleria Tower, 13155 Noel Road, Suite 1600+
Copyright © Levi Strauss & Co., Levi's Plaza, 1155 Battery Street, San Francisco,+ 121Time.com
Zara © 2010 ZARA, all rights reserved Threadless © 2011, a skinnyCorp LLC company. All designs copyright by owner
YouTube © 2011 YouTube, LLC
Wikipedia® is a registered trademark of the Wikimedia Foundation, Inc., P.O. Box 78350, San Francisco, CA 94107-8350, USA, http://wikimediafoundation.org
MINDSTORMS is a trademark of the LEGO group © 2011 The LEGO Group. All rights reserved
Myspace © MySpace, Inc. 407 N. Maple Drive, Beverly Hills, United States, 90210
Copyright © 2011 - NineSigma, Inc. All Rights Reserved
Copyright © 2011 InnoCentive, Inc. All rights reserved.
Planet Eureka © 2009–2011. Eureka! Institute. Inc. All Rights Reserved. Used under license by Merwyn Research, Inc.
Patents Granted or Pending

A part from any fair dealing for the purposes of research or private study, or criticism or review, as permitted under the Copyright, Designs and Patents Act 1988, this publication may only be reproduced, stored or transmitted, in any form or by any means, with the prior permission in writing of the publishers, or in the case of reprographic reproduction in accordance with the terms of licenses issued by the Copyright Licensing Agency. Enquiries concerning reproduction outside those terms should be sent to the publishers.
The use of registered names, trademarks, etc., in this publication does not imply, even in the absence of a specific statement, that such names are exempt from the relevant laws and regulations and therefore free for general use.
The publisher makes no representation, express or implied, with regard to the accuracy of the information contained in this book and cannot accept any legal responsibility or liability for any errors or omissions that may be made.

Printed on acid-free paper

Springer is part of Springer Science+Business Media (www.springer.com)

Preface

Today's global marketplace is a perfect storm of economic instability, fierce competition and consumer volatility that buffets demand and creates turbulence for the supply chain operations of businesses everywhere. Such a context has created a sense of impermanence for even the most meticulously planned and robust supply chain design, and is rendering uncompetitive those supply chains that are forecast-driven, unresponsive and predicated on economies of scale. The new challenge is to design and implement customer-driven supply chains that are inertia-free, attuned to mass customisation regimes, can synchronise supply with demand, and routinely manage and co-ordinate small-quantity, high-variety production and distribution.

Some of the more-recently published works on supply chain management have included a commentary or section on the notion of customer-driven or demand-driven supply chain design, and a few have recognised the potential performance gains offered by the customer-driven concept, but, in this book, we have composed a text where customer-driven supply chain design takes centre stage. Much of the material within the text will be familiar to students and researchers of supply chain management, but we have attempted to bring together several key strands to pursue the theme of customer-centricity and its implications for supply chain design and performance. Specific topics concern supply chain strategy formulation, the explanation and examination of customer-driven processes, the role of information systems in supporting customer-driven supply chains, mass customisation, horizontal partnerships and open innovation, the modelling and simulation of supply chains, and the measurement of customer-driven supply chain performance.

Chapter 1 highlights the complex anatomy of supply chains, explains core competence and emphasises the value of regarding a supply chain as an integrated whole. Such an holistic perspective provides the necessary vantage point to focus on overall supply chain performance, appreciate the value of co-operation, share a common vision with supply chain partners, and formulate a strategy that can be meaningful and coherent across a series of network alliances. Supply chain strategy should be composed of a range of sourcing, operations and logistics decisions that operate in concert and support corporate goals. It should be formulated with collaborating partners in mind in order to have the best opportunity to maximise overall supply chain performance. The immutable truth is that every supply chain is different and supply chain strategies cannot be cloned but need to be attuned to the type of product supplied and the specific market sector served. A popular approach presented and discussed within this section is to base supply chain strategy on the nature of product demand.

Chapter 2 highlights the principles of the customer-driven supply chain—planning and activity synchronisation through information sharing and customer-driven processes. Four wide-ranging ambitions—the elimination of waste, the alignment of production with demand, the integration of suppliers, and the creative involvement of the workforce in process improvement activities—are discussed to articulate the meaning of customer-driven processes.

In Chap. 3 we introduce the term 'glass pipeline' to reflect the need for information systems to support the customer-driven concept by having a commitment to sharing demand and production data across supply chain tiers so that customer and supply chain behaviour are transparent to decision makers. The 'next process is our customer' philosophy is no more important than the 'previous process was our supplier'—businesses operating in supply chains must help their 'subordinates' not just through process improvement but also through information improvement. A range of different, glass pipeline-enabling information and communication technologies are discussed. Informating sharing and glass pipelines are nothing without imagination so effective execution of the glass pipeline concept is not only dependent on the appropriate use of information and communications' technologies but also requires inter-organisational collaboration and agile processes that can take advantage of real-time access to data. Agility is predicated on responsiveness and is essential in order for companies operating in supply chains to be able to take appropriate advantage of glass pipelines. Agile processes facilitate rapid re-configuration in response to change and so provide the means to align production and supply chain activity with demand. Postponement doctrine, the use of the customer-order decoupling point as an important factor in fostering agility and the synergistic effect of consolidating information sharing with a material-flow innovation such as vendor-managed inventory are also discussed.

In recent years, mass customisation has emerged as a powerful approach for embracing customer centricity by regarding customers as individuals, supporting the proactive development of products and services according to an individual customer's preferences, and efficiently producing and distributing these offerings. In Chap. 4 we discuss the background to mass customisation and its underlying, fundamental capabilities—solution space definition, the design of robust processes, and choice navigation. A range of different cases are presented which demonstrate how customers can benefit from the provision of tailored solutions and experiences.

Embracing the customer-driven concept requires effective collaboration and the subject of collaboration is the focus of Chap. 5. Supply chain improvement activities are often confined to vertical structures and alliances but horizontal collaboration is making a growing and more conspicuous contribution to supply chain design. In the logistics sector, for example, horizontal alliances have become commonplace and are principally used to reduce transport costs, access new markets and to enhance customer service. Synchronous supply is supply chain material flow at its most sublime and manifests itself in the automotive sector in the form of synchronised sequencing. The adoption of sequenced supply with first-tier suppliers located on supplier parks adjacent to vehicle assembly facilities is a vertical innovation that is presented and discussed. The synchronisation and zero inventory advantages created by such arrangements are made possible by the proximity of the supply chain partners and the metronomic nature of the material flow. Both vertical and horizontal collaborations can be used to facilitate open innovation. The fundamental thinking behind open innovation is that new ideas are not solely created internally within organisations, but are also created in collaboration with external partners. An open approach to the process of innovation is described. This allows a business to overcome its lack of internal capabilities and competences, to share valuable resources and to reduce risk.

Chapter 6 describes how supply chains can be modelled and simulated from a business process and business data perspective. Supply chains are complex and dynamic systems and modelling frameworks and architectures can provide the means to understand their components, processes and relationships in an integrated fashion. Specific modelling techniques support the analysis and design of different perspectives helping managers to understand the whole system and design new solutions to improve supply chain performance. Simulation models support the understanding of the dynamism of complex supply chains. They can help lift the fog of supply chain transactions and evaluate the benefits derived from alternative scenario testing prior to adoption and implementation.

Chapter 7 introduces the subject of supply chain performance measurement and proposes a series of inter-organisational metrics that are compatible with the notion of customer-driven supply chain design. Four categories of measures are considered—behavioural, responsiveness, reliability and cost. Behavioural measures concern supply chain co-ordination and include the measurement of demand amplification and supply chain synchronisation. Responsiveness measures are associated with the flexibility of the pipeline of activities that constitute the supply chain and include forecast accuracy, supply chain cycle time, pipeline inventory and value-adding contribution. Reliability measures are concerned with the satisfaction of customer service and any disruptive effects of activities within the chain. Stockouts and backorders are relevant measures of reliability. Cost measures include transportation and inventory holding cost, and can be used to establish a quantitative cost profile across a chain.

Chapter 8 summarises three case studies from the automotive sector in order to demonstrate how the principles of customer-driven supply chain design in the form of synchronised planning, production and delivery activities between supply chain tiers, and the assessment and improvement of customer-driven processes can be achieved through the provision of a glass pipeline system for information sharing.

Each of the three automotive cases demonstrates how the glass pipeline concept, allied to the inculcation of a range of customer-driven guidelines and practices, can be used to facilitate a customer-driven way of working in high-energy supply chains. For each case, the prototype information architecture described makes production requirements accessible across the supply chain for suppliers to view on the Internet. The cases explain how the availability of a forward, visible workload can result in more predictable production schedules, reduced safety stocks, and improved responsiveness, co-ordination and synchronisation.

Andrew Lyons
Adrian Coronado-Mondragon
Frank Piller
Raul Poler

Acknowledgments

The research leading to this text has received funding from the European Community's Seventh Framework Programme (FP7/2007-2013) under grant agreement no. NMP2-SL-2009-229333 and from the UK's Engineering & Physical Sciences Research Council (EPSRC) via the Future Supply Innovations (FUSION) research project at the University of Liverpool.

Contents

1 The Development of Supply Chain Strategy 1
   1.1   Introduction 1
   1.2   Developing an Understanding of Supply Chain Strategy....... 2
   1.3   The Make-or-Buy Decision 7
   1.4   Generic Strategies and Competitive Criteria 10
   1.5   Innovative Versus Functional Products 13
   1.6   Product Life Cycles 16
   1.7   Chapter Summary 18
   References 18

2 Conceptualising the Customer-Driven Supply Chain 21
   2.1   Introduction 21
   2.2   Customer-Driven Processes 22
       2.2.1 The Elimination of Waste 22
       2.2.2 The Alignment of Production with Demand 35
       2.2.3 The Integration of Suppliers 41
       2.2.4 The Creative Involvement of the Workforce 42
   2.3   Chapter Summary 43
   References 44

3 Glass Pipelines: The Role of Information Systems in Supporting Customer-Driven Supply Chains 45
   3.1   Introduction 45
   3.2   Supply Chain Synchronisation........................... 46
       3.2.1 Information Sharing ('The Glass Pipeline') 46
       3.2.2 Achieving Synchronisation Through Agility 51
   3.3   Postponement and Mass Customisation 52
       3.3.1 Mass Customisation and the Customer-Order Decoupling Point 52

		3.4	Supply Chain Information and Communication Technologies	58
			3.4.1 EDI	58
			3.4.2 ERP Systems	59
			3.4.3 Web-Enabled Systems	62
			3.4.4 RFID	64
			3.4.5 Other Supply Chain Information Systems	64
		3.5	Chapter Summary	67
		References		68
4	**Mass Customisation: A Strategy for Customer-Centric Enterprises**			71
		4.1	Introduction	71
		4.2	The Development of Customer Orientation and Customer Centricity	74
		4.3	Mass Customisation: Definition and Literature Review	76
		4.4	Three Capabilities of Mass Customisation	77
			4.4.1 Solution Space Development	78
			4.4.2 Robust Process Design	83
			4.4.3 Choice Navigation	85
		4.5	A Review of Mass Customisation	89
		4.6	Chapter Summary	91
		References		91
5	**Network Collaboration: Vertical and Horizontal Partnerships**			95
		5.1	Introduction	95
		5.2	Outsourcing and Supplier Parks	96
			5.2.1 Introduction	96
			5.2.2 The Implications of Modularity	97
			5.2.3 Physical Configurations of Supplier Parks	98
			5.2.4 Proximate Supply and Vertical Integration	100
		5.3	Horizontal Partnerships	102
			5.3.1 Horizontal Collaboration in Logistics	102
			5.3.2 Open Innovation	106
		5.4	Chapter Summary	109
		References		110
6	**Modelling and Simulating Supply Chains**			113
		6.1	Introduction	113
		6.2	Supply Chain Structures	114
		6.3	Supply Chain Management	115
		6.4	Modelling Frameworks and Architectures	119
		6.5	Supply Chain Modelling Techniques	121

Contents　　　　　　　　　　　　　　　　　　　　　　　　　　　　　　　　xiii

    6.6   Supply Chain Simulation 124
    6.7   Chapter Summary 129
    References 130

7 Supply Chain Performance Measurement 133
    7.1   Introduction 133
    7.2   Measures of Supply Chain Performance 133
    7.3   Behavioural Measures 135
        7.3.1   Measuring the Bullwhip Effect 135
        7.3.2   Establishing a Synchronisation Index 138
    7.4   Responsiveness Measures 139
        7.4.1   Forecast Accuracy 139
        7.4.2   Supply Chain Cycle Time 140
        7.4.3   Pipeline Inventory 140
        7.4.4   Value-Adding Contribution 142
    7.5   Reliability Measures 142
        7.5.1   Stockouts 143
        7.5.2   Backorders 143
    7.6   Cost Measures................................... 143
        7.6.1   The Cost of Holding Inventory 143
        7.6.2   Transportation Costs......................... 144
    7.7   Value Stream Mapping Measures 145
    7.8   Chapter Summary 147
    References 148

8 Designing Customer-Driven Supply Chains: Illustrative Cases ... 149
    8.1   Introduction 149
    8.2   Case A 151
    8.3   Case B 157
    8.4   Case C 158
    8.5   Improving Supply Chain Performance Through Glass
           Pipeline Design Changes 160
        8.5.1   Glass Pipeline Trials for Case A 163
        8.5.2   Glass Pipeline Trials for Case B 165
        8.5.3   Glass Pipeline Trials for Case C 166
        8.5.4   Discussion of Glass Pipeline Trials 167
    8.6   Analysis of Customer-Driven Processes 167
    8.7   Extending Synchronisation Upstream in the Supply Chain 169
        8.7.1   Approach to Extending Synchronisation
              Beyond a First Tier.......................... 170
        8.7.2   The Impact of Implementing Second-Tier
              Sequencing in Case C 172

8.8	Chapter Summary	173
	Annex A: Output from the GP Trials	175
	Annex B: Extending Synchronisation in the Supply Chain	189
	References	192
Index		193

Chapter 1
The Development of Supply Chain Strategy

1.1 Introduction

In a highly competitive business world, it has become widely recognised that supply chains play a pivotal role in shaping competitive performance. A typical supply chain is an arrangement of component and materials' providers, production plants, distribution facilities, retailers and consumers connected by the downstream flow of materials and credit and the upstream flow of information and money.

A populist view of the structure of a supply chain is one where the relationships between the various production and service entities within the chain are represented linearly and connected by a series of bilateral interactions (Fig. 1.1), conveying goods from a supplier to a manufacturer, distributor, retailer and finally to an end consumer.

In reality, this representation is overly simplistic. Contemporary supply arrangements are more appropriately defined and represented as supply webs or networks, rather than linear chains, with multiple linkages between production and service entities as depicted in Fig. 1.2.

The overall network structure is predicated on the upstream ('inbound') supply network design and the downstream ('outbound') distribution network design. Figure 1.3 illustrates, in a linear and hence much abbreviated form, just some of the many alternative structures that occur. Referring to Fig. 1.3, the traditional supply chain commonly occurs with standardised, limited-variety products which consumers typically purchase directly from retailers and have no opportunity to customise their product choice. Examples include grocery items, household goods and toiletries.

An example of the direct sales supply chain structure is Dell's direct model (Dell 2006) where products are assembled to end-consumer order and delivered directly to consumers without the use of retailers. Personalised engraving of Apple's iPods, and distribution centres acting as fulfillment centres satisfying customers' orders directly rather than shipping goods to retail stores, as in the case of Tesco Direct (SupplyChainStandard.com 2009), are examples of distributor-postponed supply chains.

The retailer-postponed supply chain structure occurs in situations where both final consumer choice and product customisation is undertaken at the retailer. Examples include customised paint from home improvement and building products' retailers such B & Q or Home Depot, and stuffed toys for children from the Build-A-Bear Workshops.

A multitude of supply chain definitions has been proposed. A selection of some of the most quoted that highlight the network structure of supply arrangements, is transcribed in Table 1.1.

1.2 Developing an Understanding of Supply Chain Strategy

For many years the concept of meeting customers' needs was considered only at an organisational level. However, this notion is changing as effective supply chain management has allowed organisatons to positively influence business performance and create competitive advantage from their external relationships and supply chain activities. Despite the obvious yet obdurate issue of separate business unit ownership, supply networks are increasingly recognised as units of competition so the competitive environment should no longer be regarded as organisational unit against organisational unit, but as supply chain against supply chain (Christopher 2005). The consequence of this is that supply chains are increasingly being looked upon from an holistic, multi-business yet integrated perspective and it is from such a vantage point that makes feasible the development of a supply chain strategy that can be meaningful and coherent across a series of both tight and loose network alliances.

In order for any functional strategy to be successful it needs to be appropriately aligned with business or corporate-level strategy. Corporate strategy is at the epicentre of the process of organisational change. It is a formal attempt to satisfy customers' needs, facilitate business growth and out-perform competitors. Functional areas should necessarily use the specifics of corporate strategy as a starting point for their own strategy development processes but also, through a process of iteration, use their own competencies in a proactive manner as an input to the process of corporate-level strategy formulation.

This requirement for business-to-function strategic alignment is as true for the supply chain function as it is for any other but whereas for other business functions the process is primarily inwardly-focused, the nature of the strategic planning process for the supply chain function is primarily externally-focused. Figure 1.4 depicts a framework that articulates the strategy alignment and devolution process. In a multi-business organisation, corporate strategy should be devolved to appropriate business-level strategies which, in turn, are devolved to functional strategies.

1.2 Developing an Understanding of Supply Chain Strategy

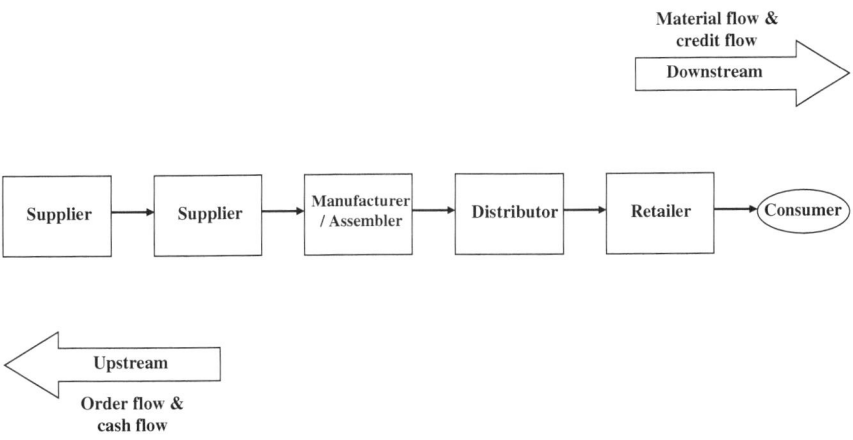

Fig. 1.1 A linear supply chain structure

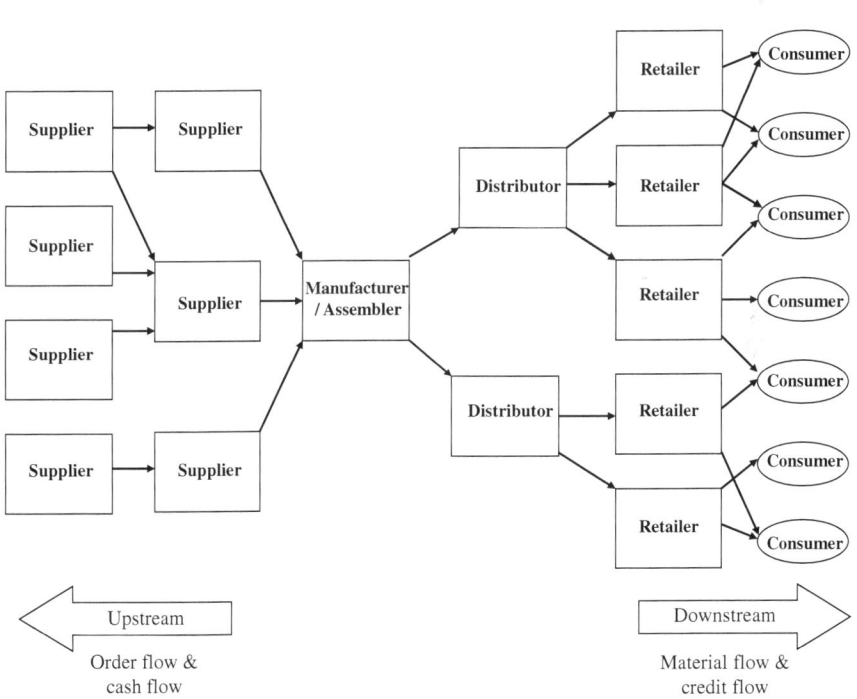

Fig. 1.2 A network of suppliers and customers

Effective supply chain strategy concerns the process of configuring a supply network in such a manner that its operation is directly supportive of corporate strategy. Effective alignment of corporate and supply strategies is essential if

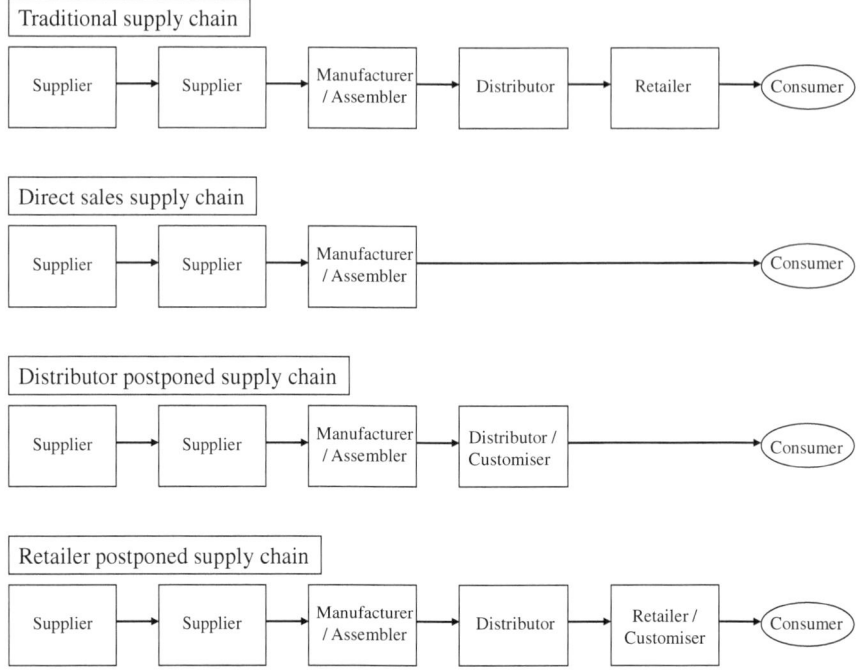

Fig. 1.3 A selection of supply chain structures

Table 1.1 A selection of definitions of the term 'supply chain'

Author	Definition
Blackstone (2008)	The global network used to deliver products and services from raw materials to end customers through an engineered flow of information, physical distribution and cash.
Gunasekaran et al. (2000)	A worldwide network of suppliers, factories, warehouses, distribution centers, and retailers through which raw materials are acquired, transformed, and delivered to customers.
Swaminathan et al. (1998)	A network of autonomous or semiautonomous business entities collectively responsible for procurement, manufacturing and distribution activities associated with one or more families of related products.
Ganeshan and Harrison (1995)	A network of facilities and distribution options that performs the functions of procurement of materials, transformation of these materials into intermediate and finished products, and the distribution of these finished products to customers. Supply chains exist in both service and manufacturing organizations, although the complexity of the chain may vary greatly from industry to industry and firm to firm.
Lee and Billington (1992)	A network of production and distribution facilities that procures raw materials, transforms these materials into intermediate and finished goods and distributes the finished goods to customers.

1.2 Developing an Understanding of Supply Chain Strategy

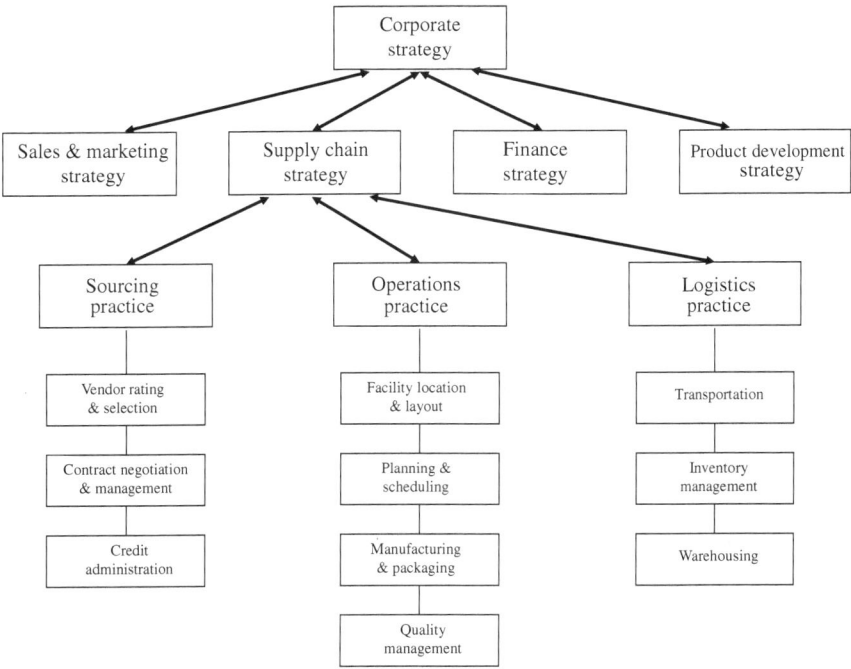

Fig. 1.4 A framework for strategy alignment

supply chains are to be designed to meet customers' needs. This is similarly the case for other business functions such as sales and marketing, finance and product development (see Fig. 1.4). The congruence between any business function activity and corporate ambition is necessary if the effective management of functional processes is to productively contribute to competitive advantage. This congruence is achieved through a process of 'strategic fit' whereby the organisation and operation of supply chain processes are aligned with customer needs through the devolution of the detail of corporate strategy. Key practices such as sourcing (vendor rating and selection, contract negotiation and management, and credit administration), operations (facility location and layout, planning and scheduling, manufacturing and packaging, and quality management) and logistics (transportation, inventory management and warehousing) need to be appropriately cohered and aligned in order to support the process of strategic fit.

Sourcing practices are primarily concerned with supplier choice and the management of the process of interaction with suppliers including the assurance of supplier performance. The planning and scheduling of resources in order to match the volume and variety challenges posed by different markets, and the physical transformation of raw materials into finished goods are the most conspicuous operations' practices. In addition, operations' practices include the location of supply network facilities, the layout of those facilities and the assurance and control of product and service quality. Logistics practices constitute the management of the physical storage and transportation of products. Inventory quantities dispersed and replenished throughout the supply network and the mode and frequency of transportation are key logistics decisions.

The volume and volatility of product demand provides necessary insight into the nature of supply chain practices and the feasible ambition that can be set for products that fall into different volume/volatility profiles. A bakery, for example, may produce a standard loaf of bread in high volume, a range of different pastries, and customised cakes for special occasions. Demand for a standard loaf is quite predictable and business ambition will be driven by a high-revenue, low-margin approach. Products will be sold from off-the-shelf and produced to replenishment orders. Production, therefore, is based on a repetitive methodology with no process changeovers, usually with highly-automated equipment, dedicated resources and relatively unskilled personnel. The product lifecycle is long so long-term relationships can be forged with suppliers and regular inbound and outbound transportation routes and schedules can be planned. Such products are fast-moving and can be regarded as 'runners'.

Pastries are driven by a desire to provide a range of different products within a single product family. Product volumes are less than that for loaves but variety, demand volatility, margins and unit costs are higher. Production is based on a batch methodology where product volumes are insufficient to justify the set-up of dedicated equipment for single products, rather fixtures and equipment are changed over each time a different product is produced. Several ingredients may be sourced from a single supplier to gain scale economies and overall product family volume can still lead to regular, predictable transportation routes and schedules. Product specifications, as with loaves, are created by the bakery—products are not customised. Products with these characteristics are termed 'repeaters'.

Customised cakes are highly differentiated products and satisfy a market niche. Volumes are low but unit costs and margins are high. Demand is volatile and a real challenge to predict. Production is to-order and based on a job-shop methodology with a very low level of repeatability, highly-skilled workers and limited automation. Specialist suppliers, that are flexible enough to respond quickly to the unique requirements of customers, are necessary. Transportation networks are intermittently used and responsive. Products such as customised cakes are termed 'strangers'.

Classification of products into runner, repeater and stranger categories allows for supply chain strategies to be developed for aggregated product families which considerably reduces the degrees of freedom in the strategy development process and improves its efficiency. Sourcing, operations and logistics practices can be based on product family characteristics and behaviours rather than individual products. Figure 1.5 provides a depiction of runner, repeater and stranger volume/volatility profiles.

1.3 The Make-or-Buy Decision

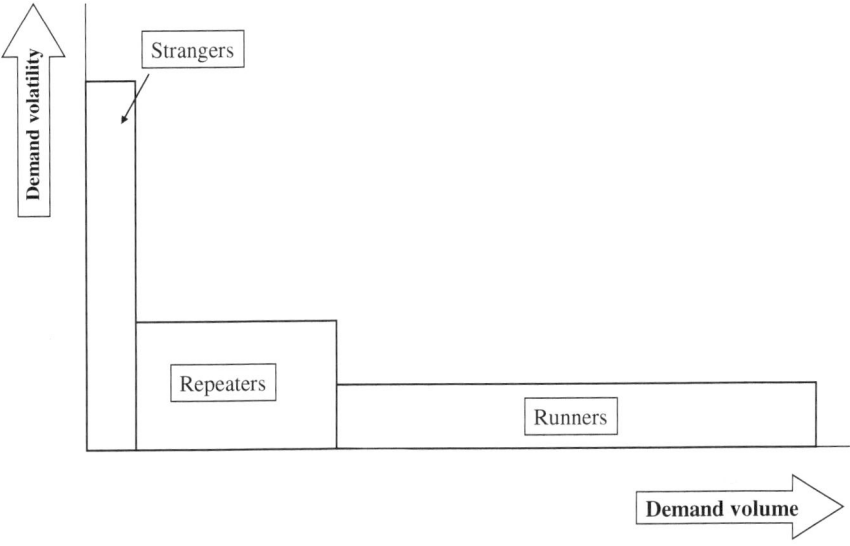

Fig. 1.5 Demand volume/volatility graph

1.3 The Make-or-Buy Decision

The make-or-buy decision articulates the business logic that distinguishes between those items that are produced internally and those that are purchased from external sources. Such a decision is an irrelevance for a retailer or distributor but for a manufacturer, the make-or-buy decision is a strategic process that is pivotal to the upstream supply network design.

A manufacturer will retain in-house manufacturing capability in order to retain what it regards as core competence and, therefore, such a decision is a critical factor for business success. Also, 'making' rather than 'buying' will be a logical decision if manufacturing costs are less than the total cost of purchasing including product transportation, or if the capability threshold of the manufacturing task precludes the involvement of third-party manufacturing, or if the available external sources are deemed unreliable. Manufacturers will relinquish manufacturing control and buy externally in order to concentrate on core competences, if manufacturing costs are greater than the total cost of purchasing and transportation or to foster a strategic partnership with a third party.

Figure 1.6 depicts an 'importance-performance' matrix that can be used to support the make-or-buy decision. The matrix is a two-by-two array where the horizontal axis depicts the item, or part, importance—an attribute that reflects its value to the business. The vertical axis depicts the manufacturing performance for the item and reflects the capability to produce the item to high standards competitively. The four quadrants are termed 'make', 'buy', 'invest or partner' and 'make or buy'. An item that is positioned in the 'make' quadrant of the matrix is

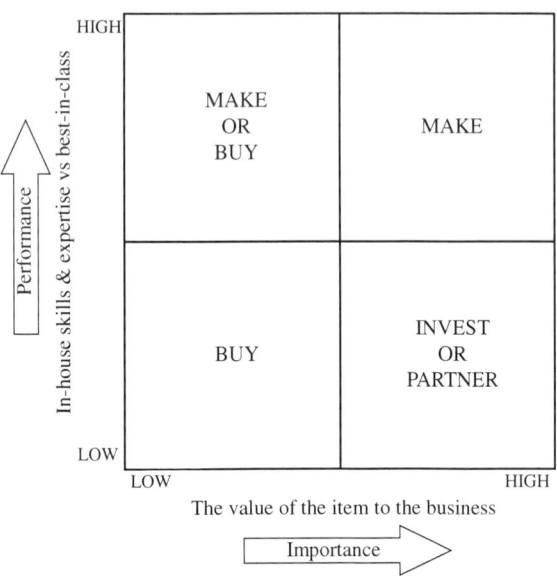

Fig. 1.6 A make-or-buy 'importance-performance' matrix

a core competence that is critical to the success of the business and is manufactured to best-in-class standards. The production of this item should be retained in-house. An item that is positioned in the 'buy' quadrant should be bought externally as it is a commodity that is produced more competitively by external companies. An item that is positioned in the 'invest or partner' quadrant is a core competence that is critical to the success of the business but is produced more competitively by external companies. In order to shift an item from this quadrant to the 'make' quadrant, an attempt should be made to make appropriate investment to improve the existing production process, or, alternatively, form an alliance with an external, world-class partner. An item that is positioned in the 'make or buy' quadrant is a non-core commodity that is produced more competitively internally than it can be by external companies. Investment should be avoided in this case but co-operation with a supplier in order to foster a relationship with a suitable external provider may lead to a future outsourcing strategy.

In a recent study (Sutter 2001) undertaken in conjunction with the authors of a make-or-buy analysis of a bumper for a new vehicle in an automotive assembly plant, a simple benchmarking study of existing internal provision against a best-in-class bumper supplier was undertaken in order to establish the performance dimension of the 'importance-performance' matrix. A summary of the results of the benchmarking study can be seen in Table 1.2.

The importance of the bumper to the business was assessed based on a standard 'core/non-core' questionnaire that established the extent to which the bumper provides a source of competitive advantage, the risk associated with outsourcing the bumper, the ownership of intellectual property and the impact on vehicle design and development capability. The subsequent importance/performance

1.3 The Make-or-Buy Decision

Table 1.2 Vehicle bumper benchmarking study

Key performance characteristic	Internal performance	Best supplier performance
Die changeover time	2.5–5 h	1–5 h
Manufacturing process time—time to produce a single bumper excluding batch and queue times	239 min	163 min
Paint shop colour changeover time	150 s	30 s
Lead time to assemble a bumper—time between the vehicle assembler's 'broadcast' of a fixed order in the system and the bumper required at the point-of-fit	1.5 h	12 h
Inventory in system—total of all inventory in the system divided by daily consumption	53 days	24 days
Length of process route—total distance each bumper travels along the production process	880 m	250 m
Value-adding time—the time the bumper spends in value-adding activities	234 min	162 min
Daily staffing level	103	58

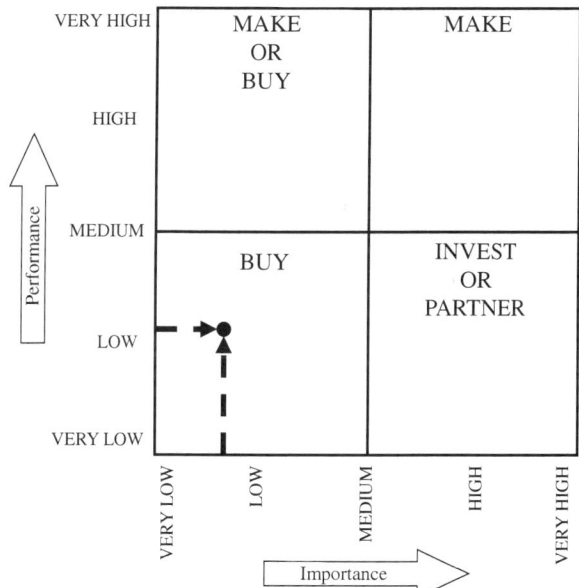

Fig. 1.7 An example 'importance-performance' matrix

matrix for the bumper can be seen in Fig. 1.7. The make-or-buy decision is determined by the point where the 'importance' line coincides with the 'performance' line. In this case, the decision was to 'buy' the bumper from an external supplier.

1.4 Generic Strategies and Competitive Criteria

The basis of competition for a business concern is often recognised as being a fundamental decision to follow one or a combination of three corporate-level generic strategies: cost leadership, differentiation or responsiveness. This is a simple extension of Porter's (1985) seminal work on competitive advantage where responsiveness is a particular form of differentiation. It is the goal of the supply chain function to be able to interpret these high-level, strategic concepts and formulate a supportive set of policies and actions. A successful execution of the strategy will allow the supply chain to positively contribute to the achievement of competitive advantage.

The adoption of a low-cost strategy requires the supply chain adopts low-cost policies both upstream and downstream. This means supplied materials are sourced primarily based on cost, modes of transportation should be chosen to be cost-effective, assets will be heavily utilised and inventory should be minimised throughout the chain.

Differentiation is principally concerned with providing a distinctive product feature or service. Supply chains can support the provision of distinctiveness through the facilitation of modular design and information sharing to handle variety and support mass customisation.

Responsiveness concerns the rapid development of new products (or improvements to existing products) and/or the fast delivery of products to customers in response to changes in customer demand. Responsiveness strategies require supply chain organisational structures have low-levels of inertia so decisions can be made quickly. In addition, other supply chain attributes that are desirable and conducive to supply chain responsiveness include co-operative supplier relationships and the sharing of information between buyers and suppliers, supply chain processes that are flexible, short lead times and a dynamic approach to capacity management. Table 1.3 summarises how supply chain decisions impact strategy.

The concept of trade-off—the recognition that a business cannot be good at everything, at least not to the same extent—is fundamental to the notion of genericity. Therefore, priortising between alternative competitive criteria is necessary. However, some further specificity and heterogeneity can be considered by sub-dividing each generic strategy. For example, responsiveness can be sub-divided into issues of flexibility, delivery speed and delivery reliability, and differentiation can be considered in terms of quality, innovative design and service experience.

A widely-cited and well-respected approach to the analysis of markets and the formulation of an operations strategy has been made by Hill (2005). A key differentiating feature, and central to the 'Hill approach', is the identification and use of order-winning and order-qualifying criteria. Hill describes order qualifiers as criteria that get a service or product into a marketplace or onto a customer's shortlist. Order winners are criteria that provide the means to be awarded business

1.4 Generic Strategies and Competitive Criteria

Table 1.3 The relationship between strategy and supply chain characteristics (adapted from Heizer and Render 2011)

	Low-cost strategy	Response strategy	Differentiation strategy
Supplier's goal	Supply at lowest possible cost	Respond quickly to changing requirements and demand to minimise stockouts	Share market research; jointly develop products and modules
Primary supplier selection criteria (Sourcing)	Select primarily for cost	Select primarily for capacity, speed and flexibility	Select primarily for product development skills
Process characteristics (Operations)	Maintain high average utilisation	Invest in excess capacity and flexible processes	Modular processes that lend themselves to mass customisation
Inventory characteristics (Logistics)	Minimise inventory throughout the chain to hold down costs	Develop responsive system, with buffer stocks positioned to ensure supply	Minimise inventory in the chain to avoid obsolescence
Lead-time characteristics (Sourcing/Operations/Logistics)	Shorten lead time without increasing costs	Invest aggressively to reduce production lead time	Invest aggressively to reduce development lead time
Product design characteristics	Maximise performance and minimise cost	Use product designs that lead to low set-up time and rapid production ramp-up	Use modular design to postpone product differentiation for as long as possible

over competitors who have qualified to be in the same market. Satisfying the 'buy' decision of customers relative to competitors is at the core of the order-qualifying and order-winning analysis. Order qualifiers need to be satisfied at a level that is comparable with competitors, whereas order winners need to be achieved at a higher level than competitors. To put another way, qualifiers get you an invitation to the ball, winners get you a dance with the belle.

Hill suggested that order qualifiers and order winners should be weighted to aid their clarification. Qualifiers are assigned to one of two classes: order qualifiers and order-losing sensitive qualifiers. The second class is used to highlight those qualifiers for which failure to satisfy "will lead to a rapid loss of business" (Hill 2005). In the case of order winners, the recommendation is to assign 100 points across all order winners within a particular market in order to establish their relative importance. Typical order winners and qualifiers include price, quality conformance, delivery speed, delivery reliability, demand increases, colour range, product/service range, design leadership, technical support supplied and new product/service time to market. Hill's framework for reflecting operations strategy issues in corporate decisions is depicted in Table 1.4.

Hill's approach uses order winners and qualifiers to underpin the development of an operations strategy that is appropriately linked to markets. Despite the

Table 1.4 Hill's framework (from Hill 2005)

Corporate Objectives	Marketing strategy	How do you qualify and win orders in the marketplace?	Operations strategy	
			Delivery system choice	Infrastructure choice
Sales revenue	Product/service markets and segments	Price	Choice of delivery system	Function support
Survival	Range	Quality conformance	Delivery system trade-offs	Operations planning and control systems
Profit	Mix	Delivery speed	Make-or-buy decisions	Quality assurance and control
Return on investment	Volumes	Delivery reliability	Capacity—size, timing, location	Systems engineering
	Standardisation versus customisation	Demand increases	Role of inventory in the delivery system	Clerical procedures
Other financial measures	Level of innovation	Colour range		Payment systems
Environmental targets	Leader versus follower alternatives	Product/service range		Work structuring
		Design leadership		
		Technical support		
		Brand name		
		New products and services—time to market		Organisational structure

1.4 Generic Strategies and Competitive Criteria

emphasis on operations rather than supply chain, Hill explicitly identifies the make-or-buy decision and the role of inventory as elements of the strategic operations' task—practices that resonate with sourcing and logistics respectively and are included in the strategic alignment framework in Fig. 1.4. Hill's fieldwork also includes several supply chain applications. An order winner, such as price, can be regarded as a focus for supply chain alignment in exactly the same manner as a generic strategy can, as depicted in Table 1.3, or the identification of a product as a runner, repeater or stranger (Fig. 1.5). An order qualifier, such as an aerospace approval for an organisation such as Airbus, is a quality and process standard a potential supplier would have to satisfy in order to be given a chance of winning an order from Airbus. Without such an approval, no order can be won.

1.5 Innovative Versus Functional Products

A landmark contribution to the development of supply chain strategy was made by Fisher (1997). Fisher classified products into two general types: "functional" and "innovative". The fundamental distinguishing factor between functional and innovative products is the nature of their demand patterns. Functional products satisfy a routine need and, as such, have relatively stable and, therefore, predictable demand, whereas innovative products, often associated with innovations in fashion or embedded product technology, have more volatile, unpredictable demand.

Each type of product should be supported by an appropriate supply chain design. Functional products are commodity items with relatively long product lifecycles and limited variety from which customers make their choice. Examples of functional products include stationery items such as document wallets, notebooks and pens. The relative predictability of the demand for these products allows for a consistent supply pipeline to satisfy demand which should, in theory, result in few stockouts. Fisher used the term "efficient" to describe the supply chain strategy that should be used to support the operations and distribution system for functional products. Efficient supply chains supporting functional products should strive to be homeostatic or constantly working to correct deviations from a designed norm in order to maintain their equilibrium.

Innovative products, by their very nature, have short lifecycles and extensive variety. Examples of innovative products include items such as mobile phones and fashion clothing. The unpredictable market reaction to an innovative product inevitably results in a higher risk of stockouts. The term "responsive" is used to describe the supply chain strategy that should be used to support the operations and distribution systems for innovative products. Responsive supply chains supporting such products should strive to be inertia-free, re-configurable, fine-tuned to the volatility of the market and constantly working to respond to changing consumer tastes.

Fig. 1.8 Matching supply chains with products (from Fisher 1997)

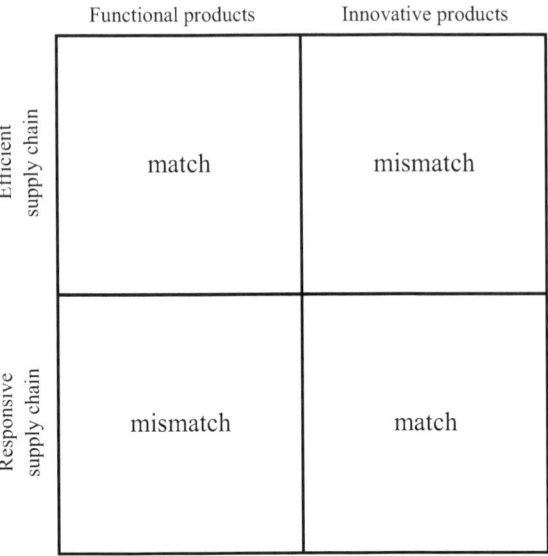

In Fisher's view, efficient supply chains should be matched with functional products and responsive supply chains should be matched with innovative products. Price is a typical order-winning criterion for a functional product and the attainment of high service levels is a typical order winner for an innovative product (Mason-Jones et al. 2000). Fisher suggested that, firstly, companies must determine whether their products are functional or innovative, hence, which products have predictable and which have unpredictable demand. The next step is for managers to decide whether their company's supply chain is physically efficient or responsive to the market. According to Fisher, companies that support either an innovative product with an efficient supply chain or a functional product with a responsive supply chain tend to be the ones with problems. Figure 1.8 depicts Fisher's framework in the form of a two-dimensional array. The horizontal axis depicts the product dimension that classifies products as either functional or innovative. The vertical axis depicts the supply chain dimension which classifies the supply chain as either responsive or efficient.

Fisher (1997) provided an explanation of the rationale for matching products and supply chains. The first consideration is that for any company with an innovative product, the rewards from investments in improving supply chain responsiveness are usually much greater than the rewards from investments in improving the supply chain's efficiency. Fisher stated in his work that for every dollar, such a company invests in increasing its supply chain responsiveness, it usually will collect a decrease of more than a dollar in the cost of stockouts and forced markdown on excess inventory that result from differences between supply and demand. Innovative products with high contribution margins cannot afford high stockout rates. High levels of responsiveness for innovative products with high

1.5 Innovative Versus Functional Products

	Innovative product	Hybrid product	Standard product
Agile supply chain	Desired application	Less desired application	Undesired application
Hybrid supply chain	Less desired application	Desired application	Less desired application
Lean supply chain	Undesired application	Less desired application	Desired application

Fig. 1.9 Matching supply chains with products (adapted from Huang et al. 2002)

margin levels more than offset the disadvantages of high labour costs. A corollary to this is that high levels of responsiveness translate into agility for the supply chain. Recent innovative products that fall in this category include Apple's iPod. Fisher explained that the economic gain from reducing stockouts and excess inventory is significant enough for intelligent investments in supply chain responsiveness to always pay for themselves. Supply chains for innovative products should be designed with reduced lead times in mind, so order deliveries take place in the shortest time. Fisher noted that "the root cause of the problems plaguing many supply chains is a mismatch between the type of product and the type of supply chain".

Huang et al. (2002), taking inspiration from Christopher and Towill (2000) and their specification and design of an agile supply chain and Naylor et al. (1999) and their exposition of "leagility", extended the Fisher matrix by introducing the notion of a hybrid product and a hybrid supply chain. Their subsequent three-by-three matrix is depicted in Fig. 1.9.

Referring to Fig. 1.9, the standard product—lean supply chain, and the innovative product—agile supply chain combinations are synonymous with Fisher's functional—efficient and innovative—responsive combinations. Efficient supply chains, which have primacy in functional products, are associated with the lean paradigm where the elimination of waste and the creation of value are top priorities. For functional products, responsiveness is as much of a key priority as it is with innovative products. However, for efficient supply chains, the elimination of excess inventory along the supply chain and the assurance of product availability are key parameters for assessing the performance of the entire supply chain. Responsive supply chains, logically matched with innovative products, are associated with the agile paradigm where the quick reaction to changing market requirements is inculcated into the supply chain's systems and processes.

The novel aspect of the matrix depicted in Fig. 1.9 is the introduction of the intermediates: the hybrid product and hybrid supply chain. Huang et al. (2002) describe the hybrid supply chain as one that is similar in meaning to Naylor et al's (1999) notion of leagility where a hybrid product that is assembled-to-order is supported by both lean and agile approaches. At the component level, the focus of the supply chain design is on being lean but at the product level, in order to adapt to changing customer requirements, the emphasis is on being agile. This hybridisation of the supply chain can support the achievement of mass customisation by postponing the point of product differentiation until the last possible moment in the chain. The approach upstream of this point is lean, the approach downstream is agile—hence the notion of "leagile". Such products often have modular product designs. This is conducive to providing consumers with wide product choice in a short lead time.

1.6 Product Life Cycles

A product life cycle (see Fig. 1.10), analogous to a human life cycle, represents a generic series of stages a product goes through from its inception through to its withdrawal. There are four stages in a typical life cycle: intoduction, growth, maturity and decline. The life cycle is often used by marketing staff to better understand sales patterns and develop marketing strategies.

Supply chain strategies also change as products move through their life cycles. In the introduction stage of the cycle, a new product or product variant is launched so, by definition, something in terms of functionality or performance that is differentiably distinct is being provided. In this stage, production run lengths are short and product demand is unpredictable so the supply chain needs to be responsive enough to ensure adequate product availability even if costs are high. If the product is complex or technically novel, supplier know-how may be a key element of the differentiating attribute. It may also be necessary for suppliers to respond quickly to quality problems and product returns. At this stage, consumer behaviour can be volatile so the supply chain needs to be operating at its most responsive level.

In the growth stage of the cycle, supply chains support an ongoing increase in the sales trajectory. Transportation networks are extended and become established

1.6 Product Life Cycles

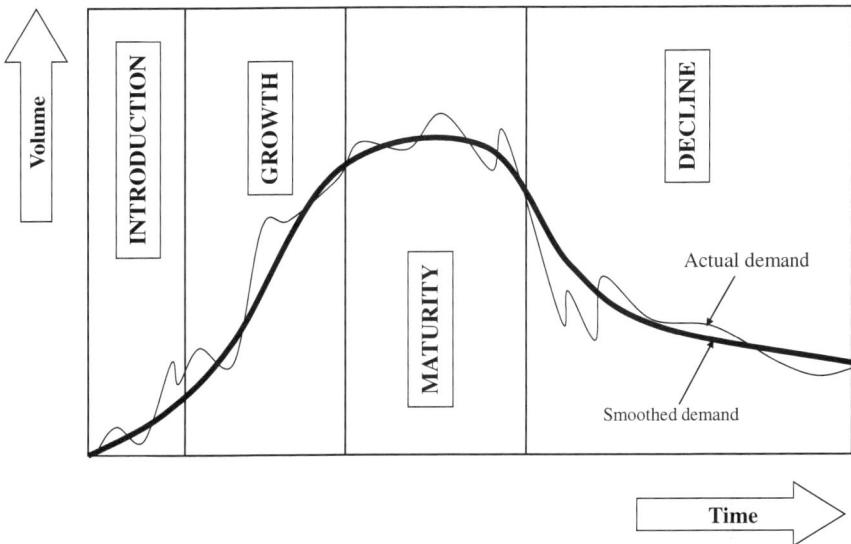

Fig. 1.10 An example product life cycle

and key partnerships between organisations are fostered and groomed. The life-cycle reaches steady state during its maturity stage where products become commoditised and demand is more predictable than previous stages. At this point, supply chain processes become standardised and inbound and outbound networks can be co-ordinated to support long production runs and high-volume sales. The supply chain should be working at its most efficient level of operation during the maturity stage. The decline stage is characterised by a reduction in capacity as the product or service is superceded by a superior alternative and sales wane.

A recent survey (Ma'aram 2010), undertaken in conjunction with the authors, of 149 companies operating in a diverse range of food supply chains in the UK and Malaysia identified and classified products that displayed lean or agile characteristics and cross-referenced this diagnosis against the stage of the product's life cycle. Food (Campbell's soup) was one of Fisher's (1997) original functional archetypes but not all food supply chains need be wholly lean in nature. Certain food items such as ice cream and customised cakes can be regarded as innovative (according to Fisher's definition). Such a viewpoint has been supported by Lee (2002). The majority of food supply chains surveyed were found to be lean but there is a greater preponderance to agile behaviour in the early stages of a product's life cycle which provides support for the received wisdom concerning changing supply chain characteristics from product inception to withdrawal. Twenty-nine companies were found to be supported by agile supply chains, of which five were found to be in the introduction stage, sixteen in the growth stage, eight in the maturity stage and none in the decline stages of their product life cycles. Lean and agile behaviours were established by asking key managers at each of the companies a number of questions concerning supply chain focus, manufacturing focus, inventory and lead-time strategies, and approach to choosing suppliers.

As a general rule, product life cycles are being compressed as competitive pressures compel organisations to continuously innovate and satisfy an increasingly discerning consumer base with new and improved products and services. Hi-tech industries, which operate within a business regime where competitive advantage is ephemeral and clockspeed (Fine 1998) is high, are particularly susceptible to unpredictable, often short product life cycles. The fluctuation of the cycle is also influenced as sales move from original equipment to original equipment plus service parts to ultimately only service parts provision.

1.7 Chapter Summary

This chapter highlighted the complex structures of contemporary supply chains and emphasised the importance of looking at a supply chain as a network of different entities that form an integrated whole. Such an holistic perspective provides the means for practising managers to focus on overall supply chain performance, value co-operation and share a common vision with supply chain partners. Supply chain goals should be set with collaborating partners in mind in order to have the opportunity to maximise overall performance across a whole product chain (or value stream). The make-or-buy decision provides the means to focus on core, supply chain competence.

Supply chain strategy differs across a product's lifecycle and is composed of a range of sourcing, operations and logistics practices that need to operate in concert and, most importantly, support corporate goals. Such goals can be expressed in terms of generic strategies, competitive criteria and product characteristics. A popular approach is to base supply chain strategy on the nature of product demand. Standard, functional products with limited variety and predictable demand should expect high-volume sales with low profit margins and should be supported by lean supply chains that minimise cost. Innovative products with high variety, volatile demand and short lifecycles should enjoy high profit margins and be supported by agile supply chains that maximise responsiveness. Hybrid supply chains combine both lean and agile characteristics and display, and support, both cost and responsiveness characteristics.

References

Blackstone JH (ed) (2008) APICS Dictionary, 12th edn. APICS, Chicago
Christopher M (2005) Logistics and supply chain management: creating value-added networks, 3rd edn. FT Prentice-Hall, New York

References

Christopher M, Towill D (2000) Supply chain migration from lean and functional to agile and customized. Supply Chain Manag Int J 5(4):206–213

Dell M (2006) Direct from dell: strategies that revolutionized an industry. Harper Collins, New York

Fine CH (1998) Clockspeed: winning industry control in the age of temporary advantage. Perseus Books, New York

Fisher ML (1997) What is the right supply chain for your product? Harv Bus Rev 75(2):105–116 March-April

Ganeshan R, Harrison TP (1995) An introduction to supply chain management. http//lcm.csa.iisc.ernet.in/scm/supply_chain_intro.html. Accesed March 2005

Gunasekaran A, Macbeth DK, Lamming R (2000) Modelling and analysis of supply chain management. J Oper Res 51:1112–1115

Heizer J, Render B (2011) Operations management. Prentice-Hall, New Jersey

Hill T (2005) Operations management. Macmillan, London

Huang SH, Uppal M, Shi J (2002) A product driven approach to manufacturing supply chain selection. Supply Chain Manag Int J 7(4):189–199

Lee HL, Billington C (1992) Managing supply chain inventory: pitfalls and opportunities. Sloan Manag Rev 33(3):65–73

Lee HL (2002) Aligning supply chain strategies with product uncertainties. Calif Manag Rev 44:105–119

Ma'aram A (2010) An examination of multi-tier strategy alignment in the food industry, PhD thesis, University of Liverpool

Mason-Jones R, Naylor B, Towill D (2000) Lean, agile or leagile? Matching your supply chain to the marketplace. Int J Prod Res 38:4061–4070

Naylor JB, Naim MM, Berry D (1999) Leagility: integrating the lean and agile manufacturing paradigms in the total supply chain. Int J Prod Econ 62:107–118

Porter ME (1985) Competitive advantage. Free Press, New York

SupplyChainStandard.com (2009) Award-winning online strategy for Tesco. www.supplychainstandard.com. Accessed December 2010

Sutter K (2001) The make-or-buy decision at an automotive plant, MSc thesis, University of Liverpool

Swaminathan JM, Smith SF, Sadeh NM (1998) Modelling supply chain dynamics: amulti-agent approach. Decis Sci 29(30):607–632

Chapter 2
Conceptualising the Customer-Driven Supply Chain

2.1 Introduction

Today's business environments are volatile and keenly contested, and even the most unfailingly-successful enterprises are finding it increasingly challenging to deliver superior returns and outperform competitors on a consistent basis; sustaining a high-level of business performance is more uncertain than it has ever been. In order to avoid any hard-earned success being short-lived, companies need to continuously transform so they are prepared for new opportunities and can confidently confront new challenges. A customer-driven enterprise embraces such continuous transformation through its capability to respond rapidly to market changes and disruptions. It effectively coheres its strategy, operations, business processes, governance structure and decision-making capabilities so that it can adapt to continually changing and unpredictable circumstances, endure disruptions to its recognised core markets and create advantages over less customer-centric competitors.

From an organisational point of view, the 'customer-driven' concept is founded on the notion of synchronisation and has two fundamental perspectives: the 'business unit' and the 'supply chain'. Customer-driven initiatives are usually concerned with analysing and improving the flexibility of the operations of a particular business unit in order to respond to changes in customer requirements. In manufacturing business units, for example, such initiatives are often employed to support the modular build of complex products. Customer-driven supply chains extend this concept to supply networks. A customer-driven supply chain has synchronised planning activities and is necessarily capable of consistently operating with small-batch, high-variety production and distribution providing customised items in single units at an efficiency level that would ordinarily be found in a mass production environment. A shift towards customer-driven supply chain management requires effective customer-driven processes and, most importantly, synchronising planning, production and delivery activities between supply chain tiers. Figure 2.1 provides a simple illustration of the customer-driven supply chain concept.

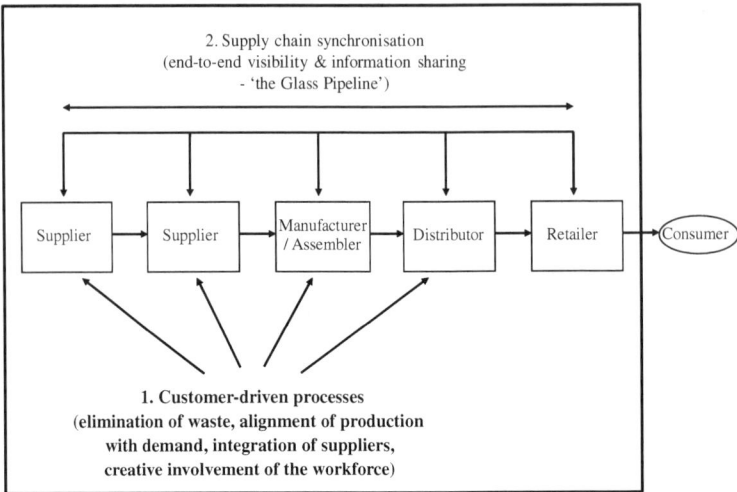

Fig. 2.1 The principles of the 'customer-driven' supply chain—planning and activity synchronisation through information sharing and customer-driven processes

2.2 Customer-Driven Processes

This chapter focuses on the exposition of the first principle—'customer-driven processes' (highlighted in Fig. 2.1). Four wide-ranging ambitions were postulated and deemed suitable by the authors to portray the sentiments of customer-driven processes (refer to Fig. 2.1):

1. the elimination of waste
2. the alignment of production with demand
3. the integration of suppliers, and
4. the creative involvement of the workforce in process improvement activities

These guidelines are distinct yet mutually supportive ideas. The alignment of production with demand, for example, is dependent on the notion of guideline '1', the elimination of waste. Common practices that can be utilised to support these guidelines and deliver customer-driven processes are depicted as part of a customer-driven process framework in Fig. 2.2.

2.2.1 The Elimination of Waste

Essentially, the elimination of waste, a principle that has had an enduring appeal as the heart of a lean approach to operations and supply chain management, strives to organise the production and service activities into an efficient configuration that

2.2 Customer-Driven Processes

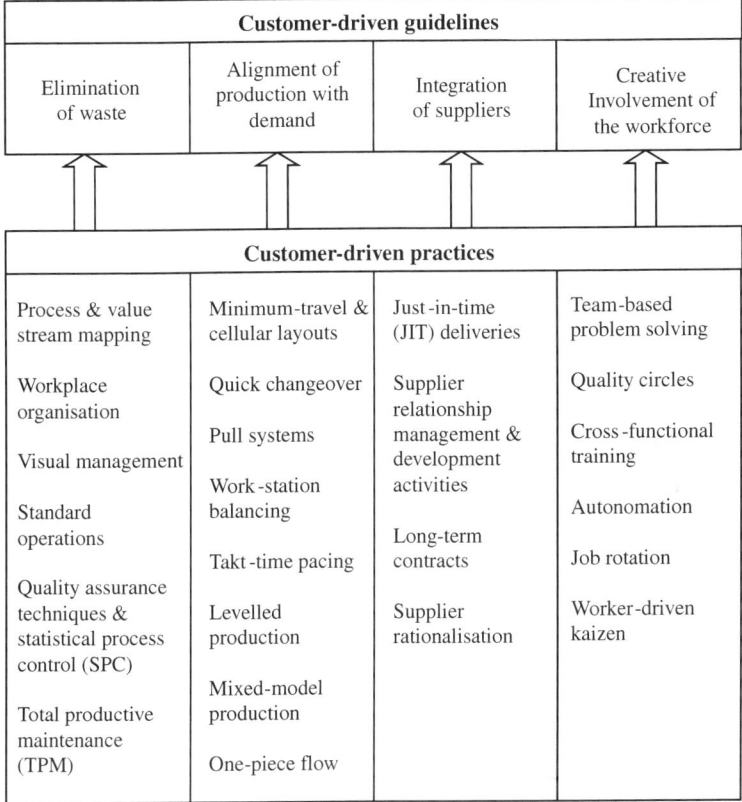

Fig. 2.2 The customer-driven process framework

facilitates an interruption-free and quality-assured flow of products. This is achieved through a structured, employee-facilitated approach to minimising and removing non-value-adding activities. Waste is regarded as any activity which does not add value to a product. Value is said to be what a customer will pay for and represents the physical processing activities that transform raw materials into component parts, components into work-in-progress (WIP), WIP into finished goods and finished goods into saleable products. There are seven, widely-recognised sources of waste:

- defects—scrap and rework add cost, extend lead times and result in poor delivery performance;
- overproduction—producing too much or too soon resulting in more inventory than is needed;
- unnecessary inventory—excessive raw materials, WIP or finished goods resulting in added costs and inflexible processes;

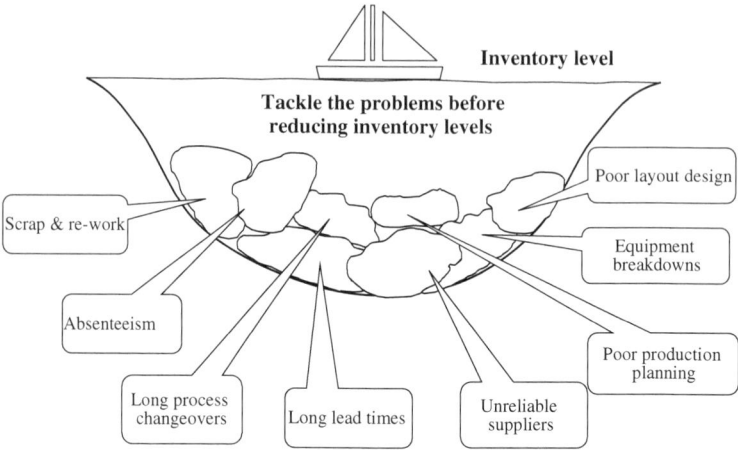

Fig. 2.3 Problems concealed by inventory

- over processing—over-complicated operations, performing more operations than required or the duplication of effort within and across organisational boundaries;
- transportation—unnecessary movement of material, or the movement of material over a longer distance than is required within or across organisational boundaries militates against small-batch production, adds cost and extends lead times;
- waiting—people unnecessarily idle within or between work tasks pass up the opportunity to be productive and add value;
- motion—unnecessary or ergonomically inappropriate movements made by people result in delays and physiological problems.

These forms of waste can create significant organisational inertia and add unnecessary cost. They arise from long machine or equipment set-ups, unreliable suppliers and logistics, unnecessary materials handling, machine and equipment breakdowns, faulty work, absenteeism, poor production planning, a lack of standardisation, poorly designed processes, inefficient facility layout, and large batch production. Figure 2.3 demonstrates how causes of waste can be concealed by high levels of inventory.

When inventory levels (represented by the water in Fig. 2.3) are high, the production and supply chain system continues operating (represented by the boat floating) and the problems in the system (represented by the rocks) are left unexposed. As the inventory level decreases, the rocks become exposed, the boat will no longer float and production ceases. The goal is to turn the rocks into pebbles and ultimately grains of sand in order for inventory levels to be reduced without disrupting production and the supply chain. As the rocks (problems) are reduced in size, the inventory can also be gradually reduced.

2.2 Customer-Driven Processes

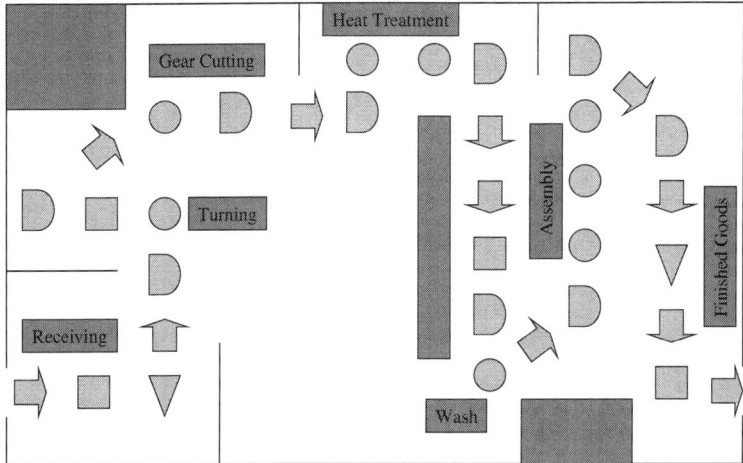

Fig. 2.4 An example process flow diagram

Mapping and charting techniques are often used to highlight waste and its causes. They provide the means to 'map-out' a production and/or supply chain arrangement by diagrammatically representing the steps involved and their interrelationships. They often use established symbols (via templates) that document and depict the flow and sequence of activities. Two pertinent techniques are process mapping and value stream mapping. Specifically, these are graphical techniques for representing and analysing processes in order to ease the identification of non-value-adding (wasteful) activities. They can also be used to enhance performance through facilitating the means to undertake targeted process improvement. A good map or chart should be simple to construct, yet sophisticated enough to aid the clear understanding of processes, unambiguous enough to communicate procedure and subtle enough to reveal waste and opportunities for improvement.

Typical process mapping constructs include a circle (○) to represent an operation, an arrow (⇒) to represent transportation, a square (□) to represent an inspection, an elongated capital letter D (D) for delay and an inverted triangle (▽) to represent storage. An operation is a physical processing activity that adds value to material, transportation represents the movement of material or a product, an inspection is an examination of the conformance to an expected measurement or quantity, a delay is the time spent in the unscheduled accumulation of material, and storage is the scheduled accumulation of material. These constructs have become commonplace in process and supply chain improvement studies. Analysts sometimes refer to the constructs as OTIDS (operation, transportation, inspection, delay, storage) symbols. A full exposition of these symbols, as part of ASME Standard 101, can be found in Barnes (1980). An example of a process flow diagram for an automotive transmissions plant is provided by Fig. 2.4.

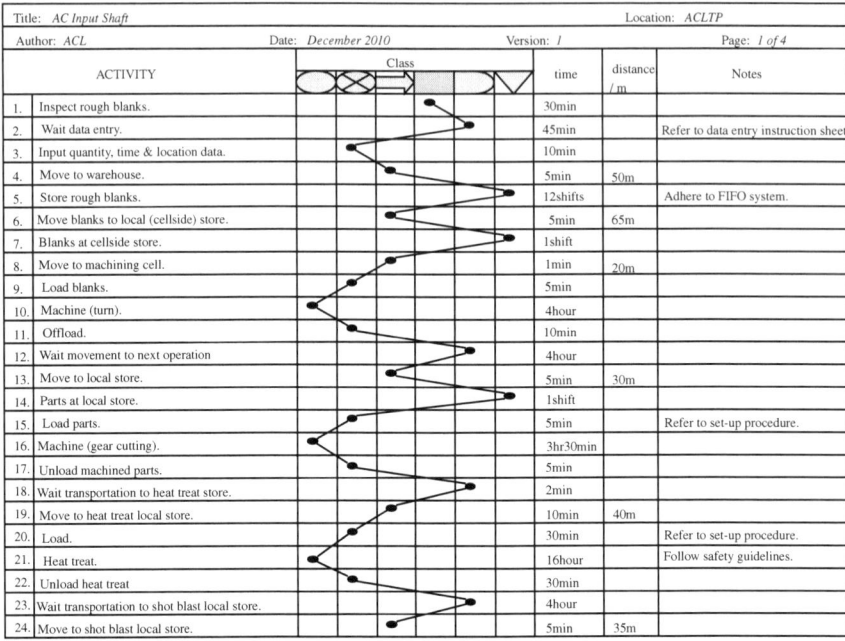

Fig. 2.5 An example process flow chart

It can be useful to add another construct to represent a non-value-adding operation, a machine changeover for example, to the OTIDS set. This can be represented by a circle with a cross through it ($\otimes$). Figure 2.5 illustrates how the symbols can be used to construct a process flow chart.

The information in Fig. 2.5 represents the first twenty-four activities in the detailed charting of the processes articulated by the flow diagram in Fig. 2.4. The chart can be used to separate value-added from non-value-added activities. This is acutely evident in Fig. 2.5 as operations have also been distinguished between those that are transformative, and genuinely value-added, such as a gear cutting process, and those that are non-value-added such as unloading the heat treatment facility. The chart can be used to interrogate current practice and improvement efforts should be focused on the incidence and magnitude of waste and underperformance. Each activity should be examined in order to consider whether it can be eliminated, reduced, simplified or combined with another activity. Sequence changes should also be considered. Activity examination often takes the form of a simple checklist. It is important to involve the owners of the process in identifying, mapping and charting working practices. Activities should be recorded accurately rather than anecdotally and it is good practice for a system of version control to be employed.

The aim of a value stream map (VSM) is to represent a complete 'value stream' for a product or family of products. VSMs are similar to process maps in that they

2.2 Customer-Driven Processes

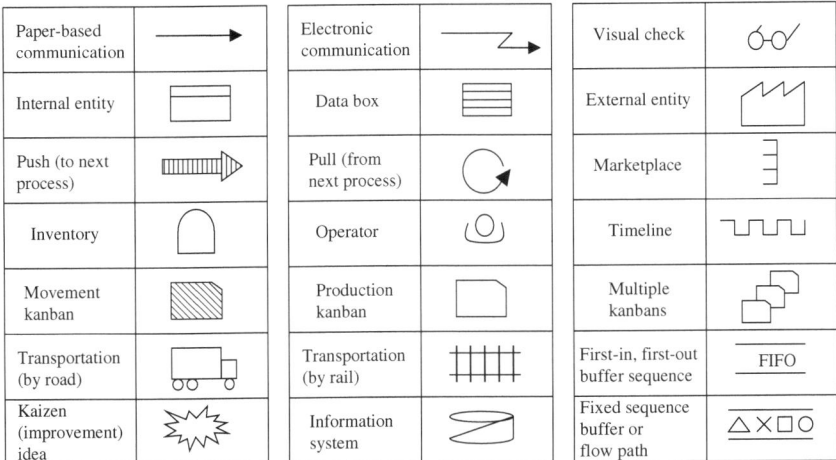

Fig. 2.6 VSM icons (adapted from Rother and Shook, 2003)

convey meaning about a production and/or supply chain system. The focus in a VSM is usually, but not necessarily broader than a process map, and the icons and constructs are capable of representing lean concepts such as kaizen and pull systems. Rother and Shook (2003) pioneered and provided a full discussion and explanation of the value stream mapping technique. Figure 2.6 is a matrix of icons that can be used in the construction of VSMs.

VSMs also enable the performance of an operation to be represented. A simple illustration of a current-state VSM for a bakery is depicted in Fig. 2.7 and its subsequent future-state VSM is depicted in Fig. 2.8. [Both figures have been adapted from an MSc research project at the University of Liverpool (Rybak 2007)]. The future-state VSM is an aspiration based on perceived improvements to the current state. Recommended changes to the current state are identified on the VSM by 'kaizen (improvement) ideas' and include the introduction of a downstream pull system, increases to the frequency of supplier deliveries and a reduction of in-process queuing. These improvements are reflected in the future-state VSM which shows an expected reduction in lead time from 15.1 days to 7.54 days.

VSMs can also be used to map supply chains. An illustration of a VSM of the multi-site production processes of a manufacturer of shower enclosures and bath screens is depicted in Fig. 2.9. It can be noticed from the VSM the significant size of the inventory buffers and the low proportion of value-adding activities. The map highlights the size of the pipeline inventory, the duration of each value-adding and non-value-adding activity and the double buffers existing between plant A and plant B. The total duration of value-adding activities for the product in question is 93.5 min (66.5 min for plant A + 27 min for plant B) while the associated pipeline inventory is 39.09 days (16.77 days + 11.18 days for plant A + 8.26 days for plant B + 2.88 days for plant C) (Coronado and Lyons 2007).

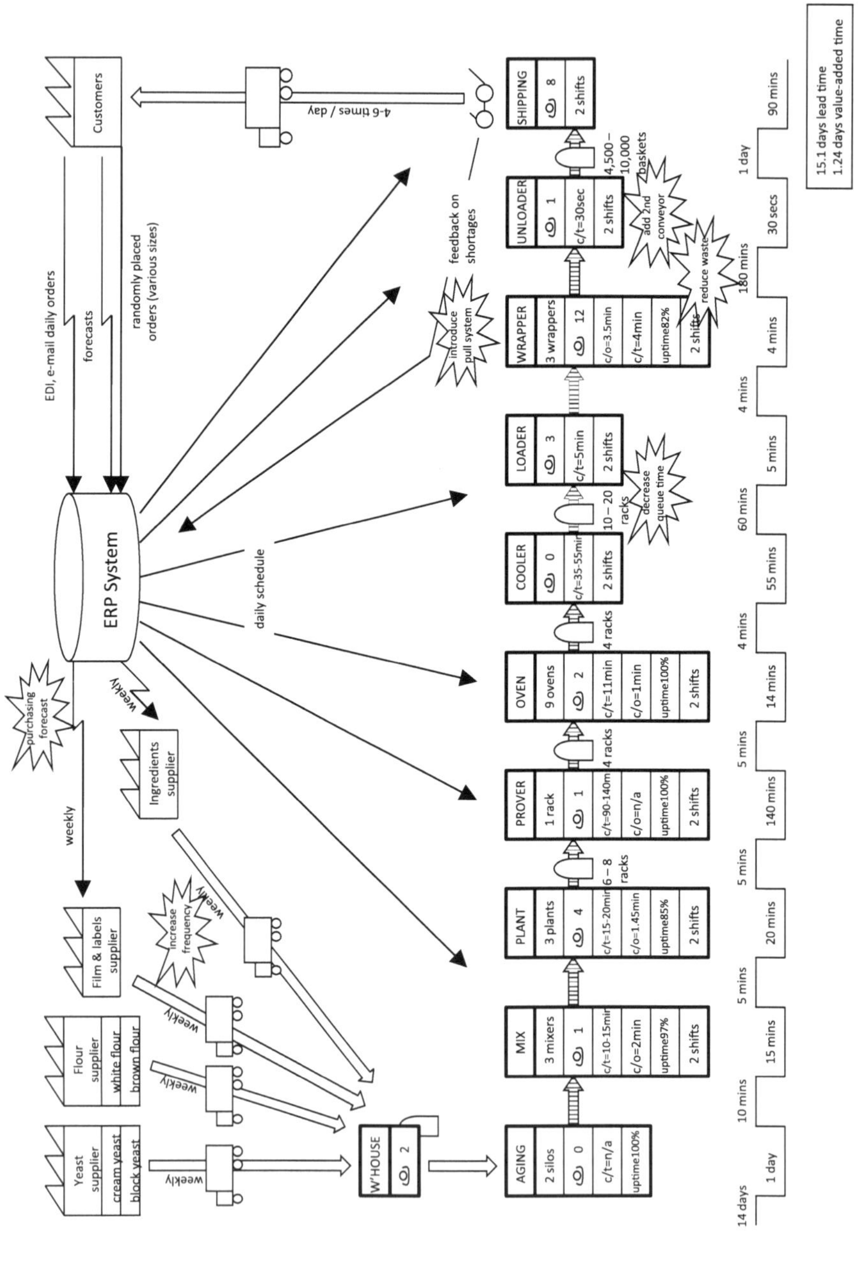

Fig. 2.7 Example current state VSM

2.2 Customer-Driven Processes

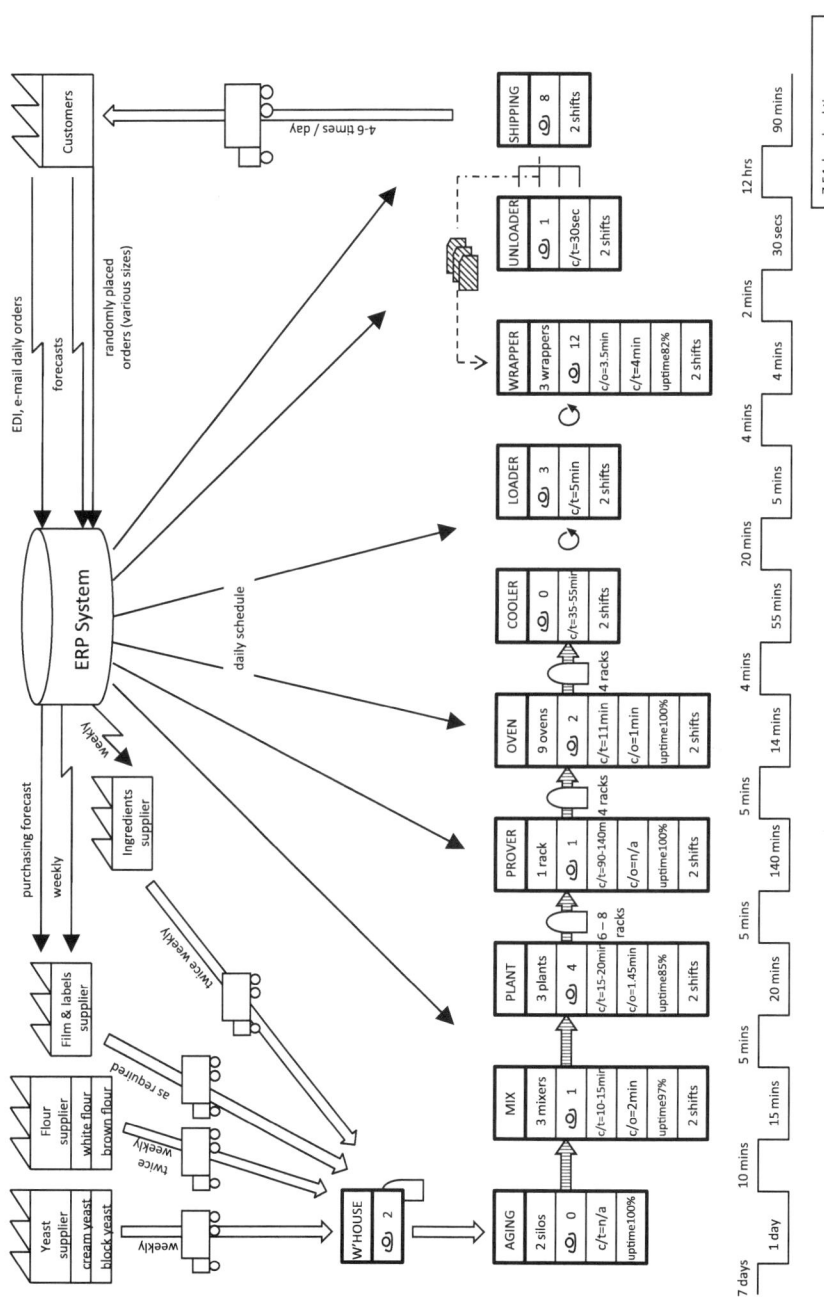

Fig. 2.8 Example future state VSM

30 2 Conceptualising the Customer-Driven Supply Chain

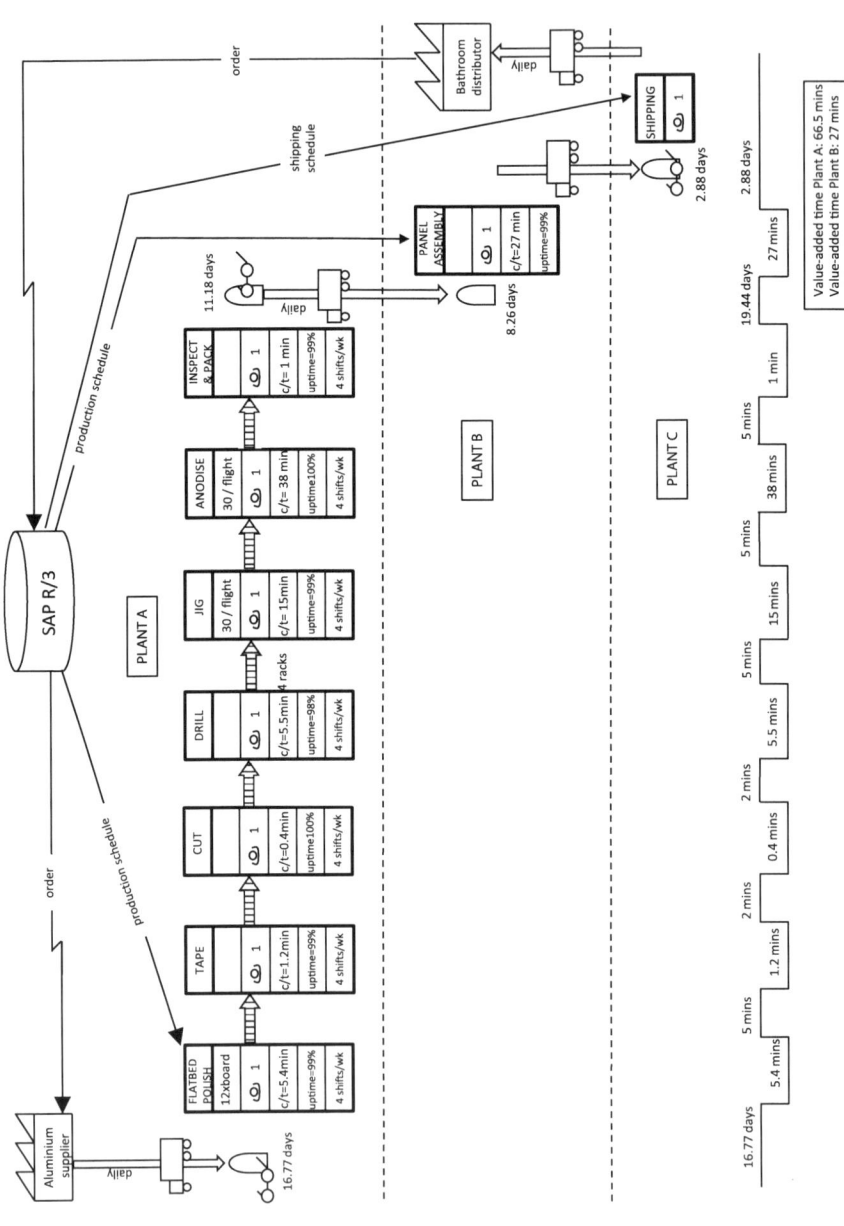

Fig. 2.9 An example VSM (adapted from Coronado and Lyons, 2007)

2.2 Customer-Driven Processes

Workplace organisation is synonymous with the 5C (or 5S) system. This is a structured procedure for establishing workplace standards for safety, cleanliness and orderliness, and for fostering a culture of continuous improvement in the organisation of an industrial or business facility. The 5C's (or 5S's) represent the first letters of each of the five stages that are undertaken to carry out the procedure:

1. Clear out (sort or seiri (in Japanese) in 5S)
2. Configure (straighten or seiton in 5S)
3. Clean and check (shine or seisō in 5S)
4. Conformity (standardise or seiketsu in 5S)
5. Custom and practice (sustain or shitsuke in 5S) (Hirano, 1995).

A pilot area is usually chosen to be the subject of a 5C exercise. This should be an area that is seen as under-performing. The team tasked with undertaking the 5C exercise is formed using people who work within the area under study in addition to outside involvement from, for example, Engineering, Maintenance and Quality Assurance staff. A 'torchbearer' for the study should be identified. Items that are considered to be of no value can be immediately removed once the 'clear out' begins. Issues and concerns should be red tagged and red tagged items are moved to a quarantine area for later scrutiny. For the 'second C' (configure), the team should examine the work area and decide where it is best to store the tools, work aids and equipment. A shadow (or outline) board provides a convenient, visually-effective method of storage. One of the key identifying characteristics of 5C is the promotion of visual organisation. Items are stored close to where they are needed and should be marked and labelled so they can be easily arranged and found. Everything should have a designated place and limits should be set for the quantities of consumables, materials and WIP. The purpose of the 'third C' (clean and check) is to determine the current condition of the machines and processes within the work area and to maintain them in good working order. Responsibilities for checking and cleaning should be assigned to appropriate staff. Regular cleaning of the work area will not only improve its operation but will also engender ownership within the workforce. The development of work standards constitutes the thrust of the 'fourth C' (conformity). These standards should be based on a consensus of best practices. Regular audits should be undertaken to ensure standards are being adhered to. The final stage (custom and practice) is to ensure that improvements that have been made and changes that have been set are systematised, become working practices that are second nature for the workforce and, most importantly, are sustained.

The effective control of production processes to assure product quality is essential to prevent defects and maintain the smooth flow of products. Processes cannot be customer-driven if they are incapable of consistently meeting specified requirements. There is variation in all things whether they are naturally occurring or manufactured. Different items produced by a process will not be identical but will vary in some way. The variation may be large or small but it will occur. For all organisations, variation can be problematic and can feasibly occur between items, within items, from process-to-process, batch-to-batch and from time-to-time. Effective quality management should ensure that process variation is controlled. There are two main classes of the causes of variation:

- Common causes are the many causes that are always present during the operation of a process. They are considered a normal part of the process. An example could be temperature or humidity. Individually, such causes may have only a small effect on the process, and can be difficult to identify and eliminate. A process which has variation due to common causes is regarded as being stable (in control) and predictable. These common causes are often termed 'unassignable'.
- Special causes are individual causes which are not a normal part of the process. An example could be an inconsistency in a supplied raw material or component. They happen only occasionally and their effect is significant enough to destabilise the process. They can frequently be traced to a single source. A goal of statistical process control (SPC) is to identify and eliminate these causes. These causes are often termed 'assignable'.

Variation in a machine or process due to common causes is said to be management controllable as it is a managerial responsibility to provide capable equipment. There is little the operator can do if the equipment is not fit for purpose and unreliable. If this is the case and equipment is not capable, quality problems will occur during production. Variation due to special causes is said to be operator controllable because the operator or other nominated person is responsible for 'keeping an eye' on the process and taking action if its condition changes. SPC helps control and reduce variation. When it is first considered, any special disturbances must be identified and eliminated, so that the process is brought to a predictable state. SPC charts are regarded as the 'voice of the process' providing information concerning the 'control' or 'out-of-control' status of the process. The control chart is used to differentiate between any common and special causes present. Fundamental statistical concepts, in particular measures of location such as mean and measures of spread such as range, help contextualise and interpret the data required for the effective application of SPC. Using established formulae, control limits can be calculated and drawn on the control chart to establish whether a process is stable and, therefore, predictable. A simplified example of a mean and range control chart is shown in Fig. 2.10

From an operational perspective, the control chart is a tool and procedure for monitoring and controlling the behaviour of the process based on rules that indicate the need for a process adjustment or intervention. Such an interruptive approach, through the use of SPC or otherwise, is an essential element of customer-driven operations. Continuous improvement (kaizen) should identify and eliminate special causes and then focus on reducing the common cause variation to provide more capable processes. As the use of SPC matures and process capability increases, the intensity with which it is used may be reduced. If a defective item is produced and identified, it should not be allowed to progress to the subsequent,

2.2 Customer-Driven Processes

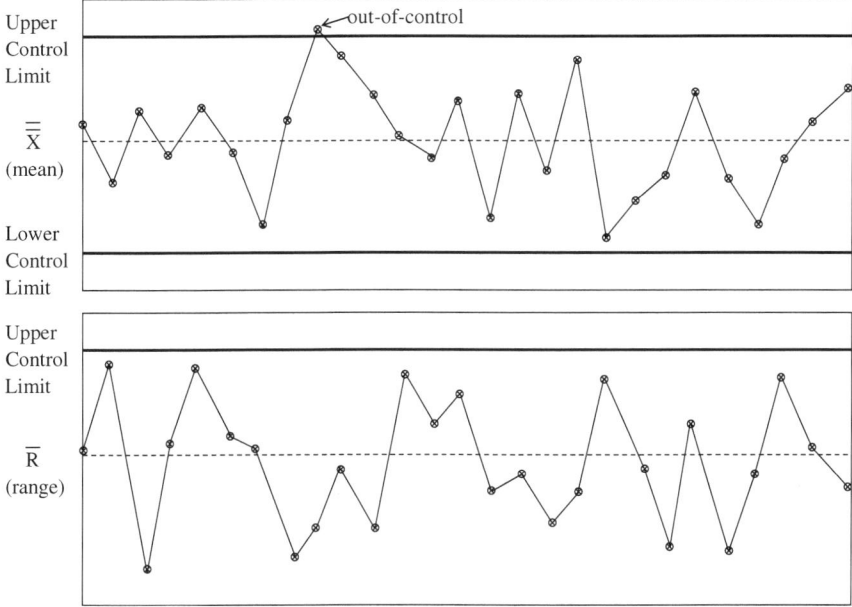

Fig. 2.10 Example SPC chart

downstream process. Halting a process as soon as a defect has been identified is known as autonomation or 'jidoka' in Japanese.

Total productive maintenance (TPM) is an approach for minimising equipment breakdowns and maximising overall equipment effectiveness within a production facility. Customer-driven operations necessarily embrace the TPM approach which aims to maximise overall equipment effectiveness by eliminating or reducing production time and volume losses in six key areas. These are known as the six major equipment losses:

- Equipment failure—caused by equipment breakdowns.
- Set-up and adjustment down time—associated with process changeovers.
- Idling and minor stoppages—delays caused by temporary product flow blockages and small deviations from normal operating conditions.
- Reduced speed—operating speeds that are less than designed or ideal operating speeds.
- Process defects—quality problems resulting in non-conforming output.
- Reduced yield—start-up losses from process start-up to steady state.

In order to maximise overall equipment effectiveness and attenuate the six major losses, formal TPM programmes advocate the use of a series of participative development activities. These activities are described in detail by the Nachi-Fujikoshi Corporation (Nachi-Fujikoshi Corporation 1990). They include:

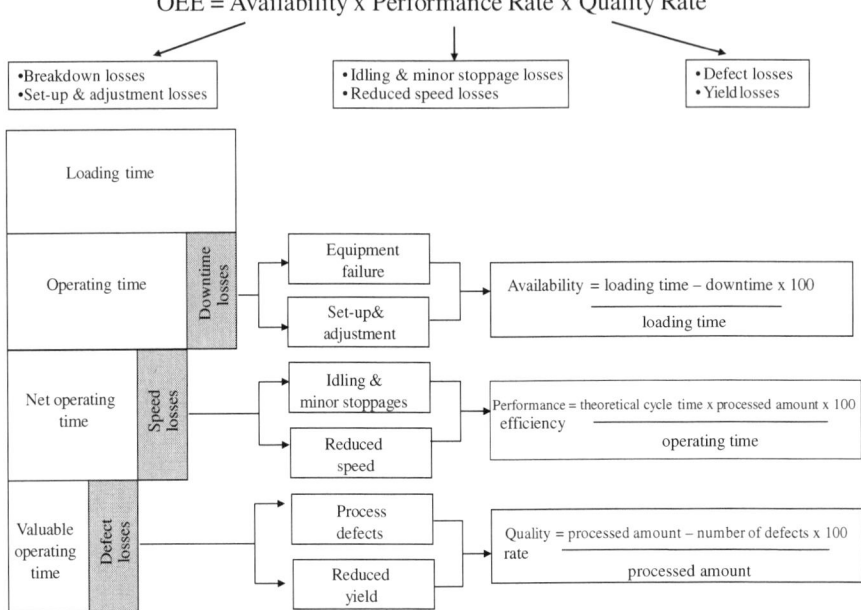

Fig. 2.11 Overall equipment effectiveness (from Shirose 1992)

- The development of autonomous maintenance practices through a structured programme of equipment care and cleaning, the creation of standard operating procedures and 5C approaches to work area organisation.
- The use of small group diagnostic and improvement activities focused on individual pieces of equipment.
- The development of a planned maintenance programme.
- An education and training programme for operators and maintenance staff.

Figure 2.11 shows how overall equipment effectiveness (OEE) can be established. OEE has become a popular measure of performance in industrial facilities and the measure is frequently used independently of the implementation of a formal TPM programme. OEE is the product of availability, performance and quality measures, and encapsulates the presence of the six big losses in a piece of equipment, or aggregated at a cell, work area or plant level.

Evidence that the principle of waste elimination is being adopted can be established from the following checklist:

- There is a real commitment to eliminate or minimise all non-value-adding activities.
- The 5C system of workplace organisation is implemented and embraced by the workforce.

- Visual displays are extensively used to support the standardisation and defect-free execution of the production process.
- Standard operating procedures are systematically used to provide work instructions.
- Abnormal process behaviour is recognised and controlled by the workforce.
- Quality systems and procedures are in place to prevent defects from moving to downstream operations.
- TPM is well established and response to equipment breakdowns is systematised.

2.2.2 The Alignment of Production with Demand

The waste elimination principle is essential to facilitate the alignment of production with customer demand. A customer-driven imperative is the ability of companies to minimise organisational inertia and configure their operations to respond to customer requirements. In customer-driven terms, the means to achieve such responsiveness is by producing at the pace of customer demand and facilitating flow through waste elimination. This combination of responsiveness and waste elimination is analogous to Spear and Bowen's (1999) analysis of the Toyota Production System and their description of its seemingly paradoxical nature in having "production flows (that) are rigidly scripted yet at the same time (having) operations that are enormously flexible and adaptable".

A fundamental component of flexible and adaptable operations is the layout of a facility. Facility layout is the arrangement of processes, work centres and service facilities relative to one another. The arrangement should facilitate the efficient operation of the production and supply chain system by minimising materials handling and by enabling a logical flow of materials and safe access and operation by the workforce. A functional layout is used in jobbing and batch environments where product (or service) volume is relatively low and variety is high. In such environments, process flexibility is typically achieved through the use of multi-skilled employees and multi-purpose equipment, and through grouping similar processing resources and equipment together. Flow paths may be variable and a general design principle is to locate those work centres with a high number of material movements between them in close proximity. A surrogate measure of materials handling can be determined using the following equation:

$$Cost = \sum_{a=1}^{n} \sum_{b=1}^{n} M_{ab} D_{ab}$$

where n = number of processes, a, b = individual processes, M = number of material movements, D = distance (assuming the same method of materials movement).

Cellular layouts attempt to gain the benefits that are ordinarily associated with repetitive environments such as high standardisation and low WIP in a jobbing or low-volume, high-variety batch environment. Cells can be created by making use of group technology, a methodology and set of principles for grouping products or parts into families based on the recognition of part/product design and production characteristics. A group technology approach may make use of part classification and coding systems to support the creation of family groupings. Cells are often discrete and self-contained and staffed by flexible, self-managed teams making use of all of the required equipment and process resources to produce a single product or a family of similar products.

A customer-driven process is designed to deliver the right items in quantities needed by downstream processes or customers at the time needed. Production strives to be demand-driven so nothing is produced until it is demanded or 'pulled' by a downstream process. A fundamental, but often obdurate, task to facilitate the alignment of production with demand is to create short, consistent process changeovers. A process changeover is an activity that takes place to make a process available to produce an item that differs in some way from the item previously produced. It is one of the six major losses of TPM and an activity that in itself is non-value added. Changeover time is the length of time it takes to change a process over from the last part of a batch to the first acceptable part of the next batch. Changeover time, therefore, typically includes the time required to change over tools and fixtures to different tools and fixtures, and make the necessary set-up and adjustments until a new part is produced that is defect free. Changeover time can be regarded as a combination of external time (when the process is still running) and internal time (when the process has stopped). Changeovers are the antithesis of the customer-driven concept as they lead to large batch production. In turn, this limits output and complicates production planning. The longer the changeover the more a detrimental impact it has on the alignment of production with demand. A structured approach to process changeovers consists of the following five steps:

1. Identify the individual elements of the changeover
2. Separate internal and external elements
3. Convert internal elements to external elements
4. Reduce internal elements
5. Reduce external elements

Step 1 of the approach concerns the separation of the composite changeover activity into its constituent specific elements. The length of time to undertake each element is measured and recorded, and the exact work content of the element is established. The actual changeover conditions are observed and, if necessary, the whole changeover is digitally recorded. Step 2 involves the separation of internal and external elements. The internal elements, undertaken when the process has stopped, such as tool and fixture removal and insertion, are established. The external elements, feasibly undertaken when the process is operating such as

2.2 Customer-Driven Processes

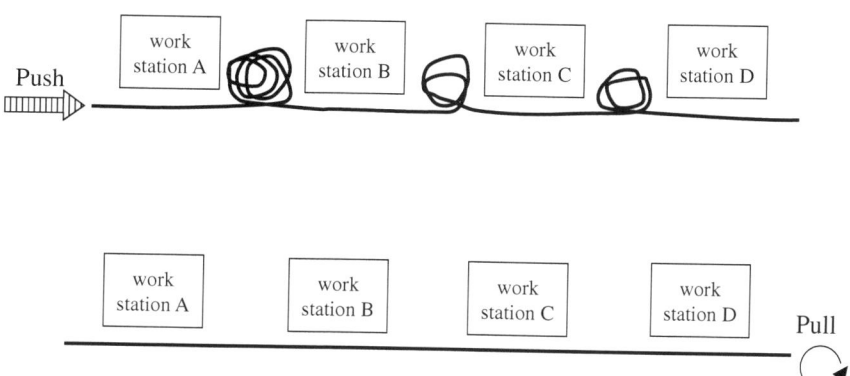

Fig. 2.12 The rope analogy

material and work aid preparation and fixture retrieval, are also identified and recorded. All activities should be classified as either internal or external. In step 3, an attempt is made to convert internal elements to external. Tooling is primed for use, fixtures duplicated and changeover schedules created. Internal elements are examined in step 4 to identify opportunities to reduce them. If possible, use is made of quick change tooling and parallel processing. A similar exercise is undertaken in step 5 to streamline external elements. Organisational issues such as workplace layout, ergonomics and the creation of schedules are addressed in this step. Documentation needed to support the changeover is created using visual management and process mapping principles. A full explanation of the process of changeover reduction can be found in Shingo (1996).

Most traditional production control systems, use what is called a 'push' methodology to move parts and material through production. A push system is based on the premise that parts and material get pushed through the operations based on a schedule. Under this type of system, an order to produce or purchase material, parts or products is released into the system at a scheduled time and is pushed from one work station to another according to that schedule.

The customer-driven concept embraces the use of a 'pull' system to move parts and materials. Instead of pushing materials through operations based on a pre-planned schedule, a pull system facilitates part movement based on the actual needs of successive work stations. This concept starts with customer demand, which pulls finished products. As the finished products are supplied to the customers, they, in turn, trigger the pull of material through the production system. Subsequent pulls work their way upstream through the production and supply chain system.

A revealing metaphor used to compare a push system with a pull system is based on representing material flow by a rope (see Fig. 2.12). In Fig. 2.12 the length of rope is analogous to the material moving through the production and supply chain system. Under the push system, coils in the rope are created at bottleneck processes and work stations. Under the operation of a pull system, the rope is not coiled, rather it is uniform and taut stretching through all processes. To move the rope through the pull system, the end is pulled; there is no need to separately manage the movement of the different coils.

Pull systems require that production is 'levelled'. Levelling refers to the production of items in a sequence where the quantities of each are produced in accordance with their demand patterns in every unit of time. This implies the production of a mixed set of items in varying quantities every hour, shift, day and week. The Productivity Development Team (1998) suggested that "levelled production allows companies to build the variety desired by customers in a smooth, mixed sequence that minimizes inventory and delays".

The use of kanbans has become synonymous with pull systems. Kanbans are work instructions and, typically, manifest themselves as visual systems where cards are used to signal the need for more materials, parts or subassemblies at a downstream operation. Ideally one item would be produced at a time. This, for many reasons, is usually impractical. For instance, the time to travel to and from one work centre to the next, or from a supplier, may be much longer than the time between requirements for the part. In these cases, it is necessary to move containers of parts rather than single units. A kanban is associated either with the movement of parts or with the production of parts. Therefore, there are two main types of kanban used, the movement (also known as the conveyance, withdrawal or supplier kanban) and the production kanban. The number of kanban cards in circulation limits the number of full containers. All of the same type of part is accounted for within the kanban system and work stations cannot accumulate extra parts outside of the kanban system. Two-card (or dual-card) kanban systems make use of separate production and movement cards and provide a higher degree of co-ordination of both production and inventory than single-card systems. Figure 2.13 is an illustration of a two card-system.

In Fig. 2.13, processes 1 and 2 have a separate upstream and downstream stock area. Each container in the downstream stock area has a production kanban attached to it and each container in the upstream stock area has a movement kanban attached to it. Some of the activities can occur concurrently and not every activity has been included but the numbers are a representation of the general activity sequence:

1. When a container of parts is removed from the downstream outbound stock area, the production kanban card is removed and placed on the production kanban board. This may be a box rather than a board.
2. The next container of parts in the downstream stock area is moved down so it is next in the queue to be consumed.
3. The movement kanban on the next container of parts in the upstream stock area is replaced with the next production kanban card on the board.
4. The production kanban work instruction is followed at process 2 and the full container is moved to the next downstream stock position.
5. The movement kanban is placed on the movement kanban board (or box).
6. The next container of parts is moved from the downstream stock area of process 1 to the upstream stock area of process 2.
7. The production kanban from the moved container is placed on the production kanban board of process 1.

2.2 Customer-Driven Processes

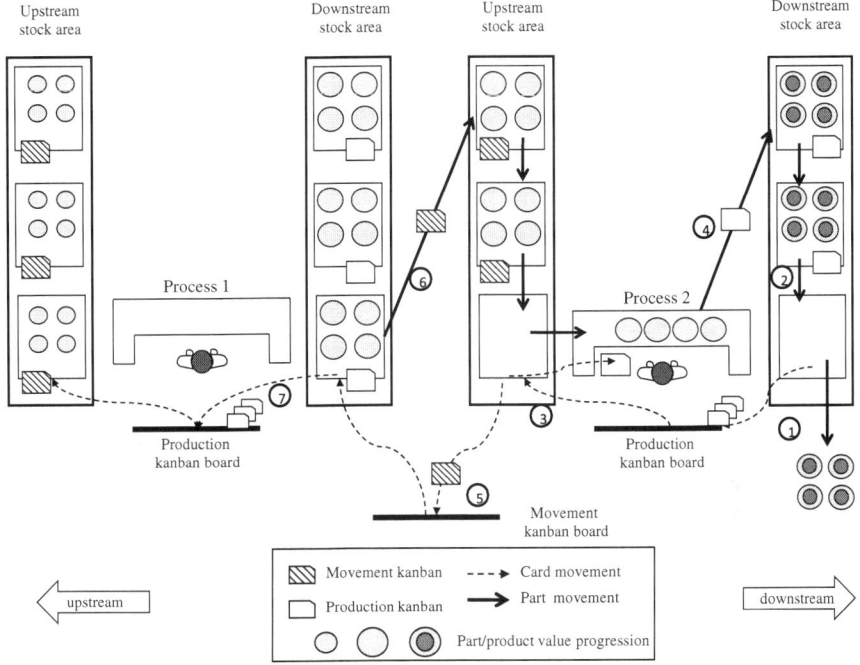

Fig. 2.13 A two-card kanban system

Product flow and the efficient allocation of resources are facilitated by the balancing of workload across work stations. The time spent at each work station should be equal, or as close as possible, to the takt time. Takt is a German word for 'rhythm' and takt time is the rhythm or rate of customer demand. It is calculated from the ratio of time available/day to demand/day. A useful visual aid to present and assess work-station balance is the yamazumi board. The yamazumi makes work, and the components of work, visible and can help identify waste and imbalance. Each work station is represented by a 'stack' of work. The stack is made up of a series of colour-coded bars sized to scale to represent time. The different colours are used to represent chunks of value-added and non-value-added work. Figures 2.14 and 2.15 illustrate examples of the use of yamazumi boards for the front seat cushion (FSC) assembly of an automotive seat assembly facility. The yamazumi board represented by Fig. 2.14 depicts the six work stations (FSC foam prep, FSC hog pass, FSC hog driver, FSC track prep pass, FSC track prep driver and FSC steam) that constitute the overall FSC assembly.

The demand is 38 seat-sets per hour equivalent to a takt time of $(60 \times 60)/38 = 94$ s. Figure 2.14 shows that the workstations are imbalanced and inefficient as each stack height is quite different to the takt time. The yamazumi represented by Fig. 2.15 is not perfect but shows an improved and more efficient balance of the FSC work stations. In this case, the tasks that had been carried out at the FSC

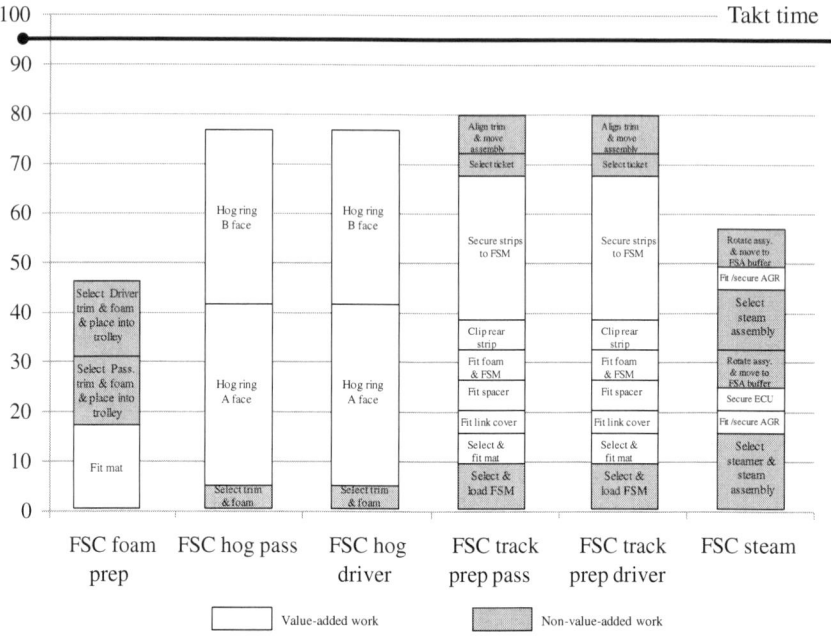

Fig. 2.14 Yamazumi example for a front seat cushion (FSC) assembly

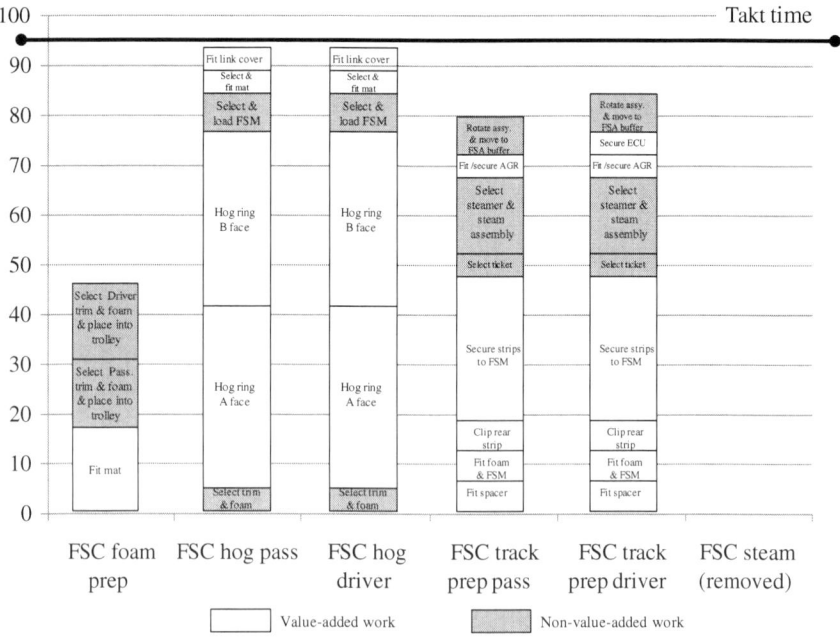

Fig. 2.15 Re-balanced Yamazumi example

2.2 Customer-Driven Processes

steam station have been allocated to the driver and passenger track prep stations. Similarly, the first three tasks allocated to the track prep stations were transferred to the corresponding driver/passenger FSC hog stations. No precedence constraints were violated in the transfer of work tasks from one work station to another.

Evidence that the principle of the alignment of production with demand is being adopted can be established from the following checklist:

- There is a commitment to reducing process set-up and changeover times.
- There is a commitment to reducing production run lengths and utilise a minimum economic batch size.
- Production is 'pulled' based upon downstream customer demand.
- Production is regarded as 'make-to-downstream requirements' rather than 'make-to-stock'.
- Production is paced to a customer demand rate or takt time.
- Production rates vary in line with customer demand rates.
- Production is mixed on the same processes and facilities.
- Changes in demand volume and mix can be easily accommodated.

2.2.3 The Integration of Suppliers

A prerequisite to the alignment concept is the third guideline, the notion of supplier integration. Without establishing a close relationship with suppliers, the concept of aligning production with demand cannot be achieved. Customer-driven suppliers need to be capable of delivering parts and materials frequently in small quantities often directly to the point of use. Supplier kanbans are treated as the same as movement kanbans in kanban-driven pull systems with external suppliers. In Fig. 2.13, supplier kanban cards would be attached to containers in upstream stock areas. When the container is removed, the supplier kanban, rather than being placed on a movement kanban board, is sent to a supplier for reorder.

Effective supplier integration is predicated upon effective supplier relationship management (SRM). Having selected an external supplier and established a contract with that supplier, it is necessary to foster and routinely assess the performance of the relationship. Relationships are the lifeblood of supply chain management. SRM is analogous to customer relationship management (CRM) where the aim is for both partners to benefit from the mutuality of the relationship. It is only through the recognition and adoption of mutually beneficial practices that are 'win–win' in nature that true integration, trust and a shared vision can be achieved. Operationally, SRM practices should facilitate appropriate communication, co-operation and co-ordination between the partners (Mettler and Rohner 2009).

Evidence that the principle of supplier integration is being adopted can be established from the following checklist:

- Suppliers are actively supported in resolving their problems and improving performance.
- Deliveries are based upon production requirements, are not excessive and arrive just-in-time.
- The notion of being a part of a complete (supply chain) value stream is both understood and accepted.
- Suppliers receive schedules that are stable and predictable without unexpected changes.
- Raw materials and ingredients are single sourced.
- Suppliers have flexible processes that can easily accommodate demand changes.
- Deliveries are made directly to the point-of-use (rather than to a remote storage area).
- Supply inventory buffers are planned and set at minimum acceptable levels.

2.2.4 The Creative Involvement of the Workforce

The inculcation of creatively being involved in process improvement activities into the minds of the workforce is essential for the elimination of waste. Total employee involvement through attempting to release the talents and creativity of people is a ubiquitous driver and ambition of process improvement approaches. For example, the Canon Production System (Campbell 1987) was founded on creative workforce involvement. Forza (1996) suggested that lean manufacturing plants are characterised by the use of small-team problem solving, worker suggestions for improvement, decentralisation of authority, use of multi-functional employees, contact between the workforce, and commitment to continuous quality improvement.

Successful performance demands that people who are responsible for the management and operation of the process are involved in the identification and implementation of process changes. The notion of harnessing the skills and experience of the 'people who know' and providing an environment in which people are comfortable in contributing to the decision-making process is essential for having a customer-driven process and should permeate through every supply chain tier. Kaizen requires a co-operative culture and is synonymous with continuous improvement through workforce engagement and embodies the notion of creatively involving the workforce in process improvement activities. The process improvement activities are usually small group activities that lead to incremental changes. Incremental rather than step changes are more likely to be sustainable and stick. Step changes are higher risk as they are more organisationally disruptive and, as a consequence, can be ephemeral. Some kaizen strategies do merge incremental and step changes but the essential thrust of customer-driven improvements is to make small progressive changes that focus on the elimination of waste, the alignment of production with demand or the integration of suppliers which together build to make significant impacts that are sustainable. The kaizen process is often implemented using the 'plan-do-check-act' (PDCA) cycle. 'Plan' activities involve identifying the problem, articulating the current state and setting objectives and targets for the kaizen team. 'Do' concerns the collection of relevant data, the use of structured approaches to the clarification of the problem causes, the generation of ideas to contain and resolve the problem and the implementation of the improvement approach. 'Check' concerns the evaluation of the results and 'Act' involves a review of the implementation and the introduction of any further changes to ensure satisfactory and sustainable performance.

Formal procedures adopted by kaizen teams to highlight and resolve problems often concern the undertaking of a structured series of steps that include team formation, the articulation of the problem, its severity and impact, the implementation of immediate containment actions, the investigation and verification of the problem cause(s) and the implementation of permanent corrective actions.

Evidence that the principle of creative workforce involvement is being adopted can be established from the following checklist:

- The workforce is actively involved in improvement activities and is empowered to make changes.
- The work environment is organised so that most work is undertaken in teams.
- Individual and team-based improvement ideas are regularly received from the workforce.
- A structured programme of employee training is in place and adhered to.
- Management devolve work-related decisions to the workforce.
- The workforce is multi-skilled and a system of job rotation is employed.
- The work culture means that change is readily accepted and regarded as the norm.
- Kaizen and constant, incremental improvement and innovation are embraced and practised by the workforce.

2.3 Chapter Summary

This chapter highlighted the principles of the 'customer-driven' supply chain-planning and activity synchronisation through information sharing and customer-driven processes. The chapter focused on 'customer-driven processes'. Four wide-ranging ambitions—the elimination of waste, the alignment of production with demand, the integration of suppliers, and the creative involvement of the workforce in process improvement activities—were discussed to articulate the meaning of customer-driven processes.

Key operations and supply chain practices associated with the elimination of waste concern process and value stream mapping, workplace organisation, visual management, standard operations, quality assurance and total productive maintenance. Key practices associated with the alignment of production with demand include cellular manufacturing, quick changeovers, pull systems, work station balancing and takt time pacing. Supplier relationship management (SRM) is an essential pre-requisite to effective supplier integration. SRM practices should facilitate appropriate communication, co-operation and co-ordination between supply chain partners so that organisational boundaries become less rigid and more porous. The creative involvement of the workforce is an attempt to release the talents and creativity of people for process improvement and productive change. Harnessing the skills and experience of the 'people who know' and fostering a culture in which people are comfortable in contributing to decision-making is essential for having a customer-driven process and should be a conspicuous part of the approach to kaizen at every supply chain tier.

References

Barnes RM (1980) Motion and time study: design and measurement of work. Wiley, New York
Campbell AT (1987) Canon production system: creative involvement of the total workforce. Productivity Press, Cambridge, MA
Coronado AE, Lyons AC (2007) Evaluating operations flexibility in industrial supply chains to support build-to-order initiatives. Business Process Manage J 13(4):572–587
Forza C (1996) Work organisation in lean production and traditional plants: what are the differences? Int J Opera Produc Manage 16(2):42–62
Hirano H (1995) 5 Pillars of the visual workplace. Productivity Press, Oregon
Mettler T, Rohner P (2009) Supplier relationship management: a case study in the context of health care. J Theor Appl Electron Commer Res 4(3):58–71
Nachi-Fujikoshi Corporation (1990) Training for TPM: a manufacturing success story. Productivity Press, Cambridge
Team Productivity Development (1998) Just-in-time for operators. Productivity Press, Portland, Oregon
Rother M, Shook J (2003) Learning to see: value stream mapping to add value and eliminate muda. Lean Enterprise Institute, Cambridge
Rybak M (2007) Operations review and production process improvements in a bakery. Msc final project report, University of Liverpool
Shingo S (1996) Quick changeover for operators: the ISMED system. Productivity Press, Portland, Oregon
Shirose K (1992) TPM for operators. Productivity Press, Cambridge
Spear SJ, Bowen HK (1999) Decoding the DNA of the toyota production system. Harvard Business Review, September/October, 97–106

Chapter 3
Glass Pipelines: The Role of Information Systems in Supporting Customer-Driven Supply Chains

3.1 Introduction

Change is the inevitable consequence of an organisation's innate customer-centricity so embracing the customer-driven concept necessarily requires a flexible approach to the management and co-ordination of those organisational behaviours, systems and processes that intrinsically favour the status quo over re-configuration and renewal. Customer-driven supply chains cannot be stolid and inert but require co-operative behaviours, flexible systems, and efficient and synchronised processes. They are also characterised by a high degree of information and material (physical) flow innovation. Innovative information systems are necessary for the sharing of customer order information throughout supply chains and innovative material flow systems are necessary for the production, handling, storage and transportation of materials and products. Innovative information systems are necessarily composed of those communication technologies and information management practices that allow data to be collected, transmitted, made accessible and conveniently stored across multi-tier supply chains so that decentralised activities can be co-ordinated across the tiers. Innovative material flow systems must allow for physical processes to be customer-driven producing and transporting materials and products responsively, at the required time and in the required amounts. Figure 3.1 charts a hypothesised evolution of the customer-driven, supply chain concept.

In Chap. 2, it was posited that customer-driven supply chain management requires effective customer-driven processes and the synchronisation of planning, production and delivery activities between supply chain tiers. Figure 3.2, identical to Fig. 2.1 but with 'supply chain synchronisation' highlighted, reiterates the illustration of the customer-driven supply chain concept.

3.2 Supply Chain Synchronisation

This chapter focuses on the exposition of the second principle—'supply chain synchronisation'. Two wide-ranging ambitions were postulated and deemed suitable by the authors to portray the sentiments of supply chain synchronisation:

1. Information sharing, and
2. achieving synchronisation through agility.

3.2.1 Information Sharing ('The Glass Pipeline')

Information flow is as important as material flow, and information management is as important as the physical processes that take place in, and often singularly characterise supply chains. The availability of information is critical for supply chain replenishment decisions, resource planning, logistics and invoicing. The flow of information along a supply chain constitutes a principal element of supply chain design and ultimately has a significant impact on the performance of the entire chain. In many industries, information is communicated only between contiguous and immediate tiers of customers and suppliers in a time-phased, sequential manner; rarely is information made accessible to tiers which are not contiguous. There are exceptions. One concerns Xilinx's Virtual Direct Program (re-named Xilinx Direct Fulfillment (XDF)) with Cisco which contravenes the conventional supply chain information flow by bypassing the usual distribution channel in transmitting replenishment information from Cisco (tier n) to Xilinx (tier n + 2) (Souza 2003).

Information sharing in supply chains has been recognised as a significant contributor to the mitigation of demand visibility problems such as excess inventory levels and poor supply chain co-ordination resulting in demand amplification (the 'bullwhip effect'). The notion of demand amplification was first recognised by, and has its origins in the work of Forrester (1958, 1961), hence the phenomenon is often referred to as the Forrester Effect. Through the use of simulation models, Forrester demonstrated how small variations in final customer demand may be amplified upstream in a supply chain. Figure 3.3 uses the same linear supply chain structure as Fig. 1.1 in Chap. 1 in order to provide a simple graphical illustration of demand amplification. In the figure, small variations in demand at the retailer are progressively amplified as orders are transmitted upstream. The most severe order fluctuations are evident at the second-tier supplier.

Since Forrester's landmark work, numerous other authors have extended the research in this area. The "beer distribution game", developed at the Sloan School of Management, is a role-playing simulation of an industrial production and distribution system that demonstrates demand amplification across a four-tier chain. It shows how the supply chain structure, the irrational behaviour of individuals and

3.2 Supply Chain Synchronisation

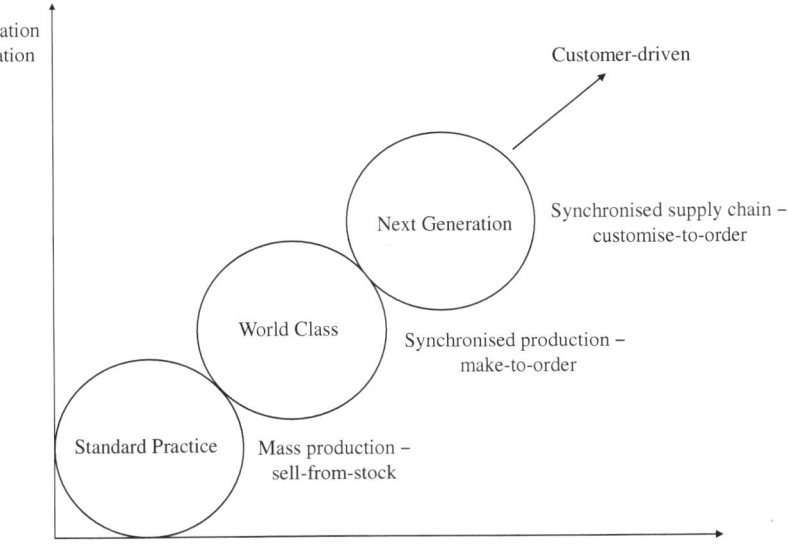

Fig. 3.1 The 'customer-driven' evolution

their "misperceptions of feedback" can lead to the generation of highly variable orders between tiers despite constant consumer demand (Sterman 1989). Sterman (1989) noted that individuals were insensitive to feedback and had a strong tendency to attribute behaviour to external factors rather than their own decisions.

Other researchers have expanded Forrester's work and tested various methods for reducing demand amplification using mathematical models, computer-based simulations and case studies. Lee et al. (1997) noted that Procter & Gamble (P & G) first coined the term 'bullwhip effect' after noticing that whereas their customers' demand for Pampers diapers were relatively steady, the demand order variabilities in the supply chain were amplified as they moved upstream. The authors highlighted a similar phenomenon at Hewlett-Packard where seemingly anomalous order fluctuations were apparent upstream of resellers (Lee et al. 1997). Lee et al. (1997) suggested four major causes of bullwhip:

1. Demand forecast updating—forecasting at each tier is based on orders received rather than actual demand distorting the upstream perception of demand.
2. Order batching—periodic ordering practices that aggregate requirements and delay the order decision amplify variability.
3. Price fluctuations—forward buying in order to take advantage of low prices or quantity discounts leads to irregular ordering patterns.
4. Rationing and shortage gaming—in situations where demand exceeds supply, customers are sometimes tempted to exaggerate their real needs when they order. Such a practice can mask the real demand picture.

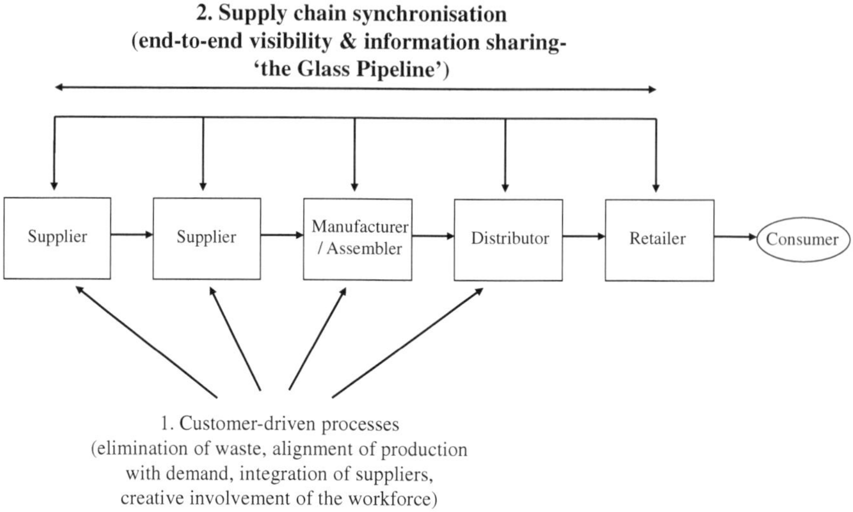

2. Supply chain synchronisation
(end-to-end visibility & information sharing-
'the Glass Pipeline')
Supplier
Supplier
Manufacturer / Assembler
Distributor
Retailer
Consumer
1. Customer-driven processes
(elimination of waste, alignment of production with demand, integration of suppliers, creative involvement of the workforce)

Fig. 3.2 The principles of the 'customer-driven' supply chain—planning and activity synchronisation through information sharing and customer-driven processes

SUPPLIER
SUPPLIER
MANUFACTURER
DISTRIBUTOR
RETAILER
CONSUMER
orders
orders
orders
orders
demand
time
time
time
time
time
First-tier supplier orders to second-tier supplier
Manufacturer orders to first-tier supplier
Distributor orders to manufacturer
Retailer orders to distributor
Consumer demand at retailer

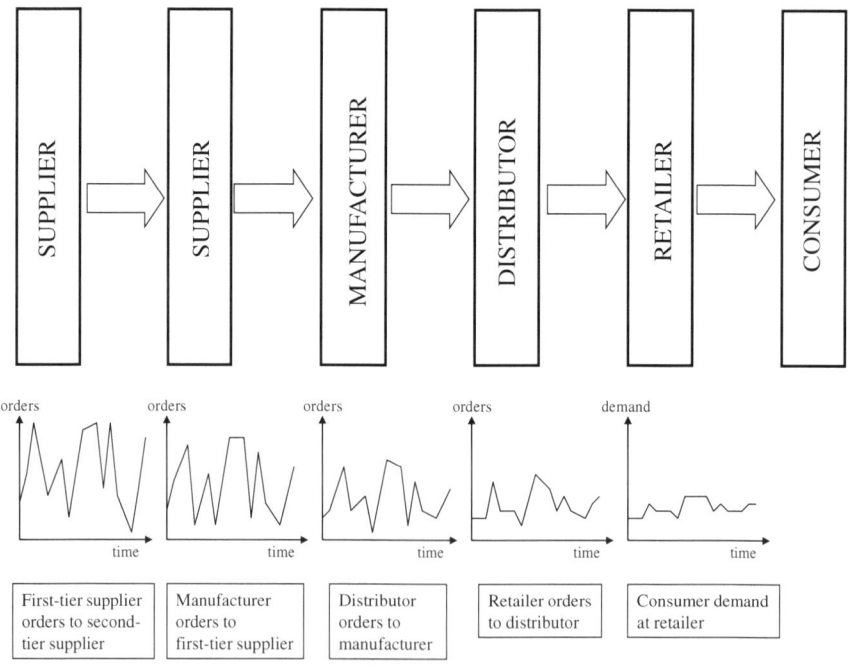

Fig. 3.3 Demand amplification

3.2 Supply Chain Synchronisation

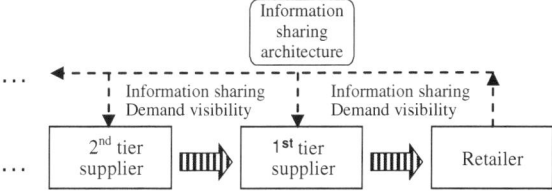

Fig. 3.4 The representation of information sharing in the supply chain

Many solutions and improvement approaches have been envisioned for the bullwhip effect. Amongst the most prominent are the exchange of sales and inventory data, the reduction of lead times and batch sizes, and increases in delivery frequency. Also it has been shown that the bullwhip effect can be markedly reduced by limiting the number and scale of promotional campaigns, using time-based quantity discounts and establishing allocation policies based on actual sales rather than orders. Drilling down even further, 'information sharing' has been recognised as a key contributor to bullwhip reduction and better supply chain co-ordination by a scholarly canon that includes Lee et al. (1997), Chen et al. (2000), Disney et al. (2004), Dejonckheere et al. (2004), Croson and Donohue (2005), Simchi-Levi et al. (2007) and Moyaux et al. (2007).

The importance given to sharing information in the supply chain is depicted in Fig. 3.4. In this highly-simplified scenario, the supply chain information architecture facilitates demand visibility and information sharing along the chain. The authors have used the term 'glass pipeline' for this concept.

Changes to the design of information flow are possible through the use of modern IT solutions. Web services, advanced software and wireless applications can support the flow of information to all tiers in a supply chain. Applications such as electronic data interchange (EDI) and the Internet have become essential to the operation of supply chains in many sectors. A glass pipeline, however, does not have a prescriptive information architecture, it is a concept, a commitment to sharing data and information across supply chain tiers so that customer and supply chain behaviour is transparent to decision makers. Demand information is made available to the immediate sub-tier and then transmitted in an appropriate form to further upstream tiers, or, if advantageous and possible, the demand information is disintermediated and made available to all, or selected tiers. Effective execution of the glass pipeline concept requires inter-organisational collaboration and trust, porous organisational boundaries and customer-driven, agile processes that can take advantage of real-time access to information.

The P & G 'solution' is based on an information sharing (glass pipeline) design. The fundamental supply chain structure for this solution can be seen in Fig. 3.5. An explanation of P & G's supply chain arrangement can be found in Lee et al. (1997) and Patton (2005).

Figure 3.5 depicts how P & G is not only directly connected with customers through the exchange of point-of-sale (POS) data but also provides aggregated POS data to suppliers. P & G has created an integrated chain that is customer-

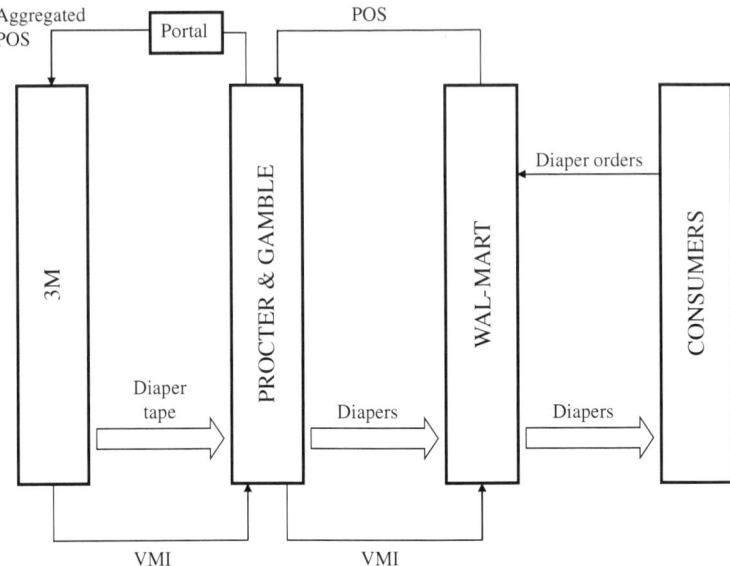

Fig. 3.5 The P & G bullwhip solution

driven through the sharing and cascading of sales data. Supporting this approach is a policy of vendor-managed inventory (VMI) between P & G and Wal-Mart and between 3M and P & G.

VMI is regarded as an alternative to the conventional order-replenishment process (Kaipia et al. 2002). The fundamental change is that the ordering phase of the process undertaken by the downstream partner is made redundant, and the supplier (upstream partner) is given both the authority and responsibility to administer the entire replenishment process. In such a manner, the replenishment decision is assigned to a single point in the chain. When using VMI, the downstream partner provides the supplier with access to inventory movements and demand/sales data and sets targets for inventory availability. The aim is to improve service and co-ordination while reducing supply chain costs. This combined approach to sharing data and utilising VMI is an example of the innovative use of both information and material flow systems, referred to at the beginning of this chapter, that typically characterise a customer-driven supply chain. Disney and Towill (2003) developed a simulation model to compare the expected performance of a VMI supply chain with a traditional supply chain and found the VMI-supported chain to better handle volatile changes in demand.

Highlighting Wal-Mart's exemplar supply chain management practices and use of VMI, Schonberger (Schonberger 2006) noted that suppliers are responsible for managing Wal-Mart's inventory replenishment via electronic access to the retailer's POS data. Suppliers ship directly to Wal-Mart under a continuous replenishment agreement instead of utilising intermittent batch shipping. The author highlighted that Wal-Mart has been able to significantly reduce flow times through distribution centres that adopt a policy of cross-docking. Cross-docked items, offloaded from an incoming trailer do not go to storage but, rather, are moved directly to trucks at other docks for immediate transfer to retail stores. Such a practice reduces lead times, inventory and warehousing costs.

3.2.2 Achieving Synchronisation Through Agility

Agility is the by-word for responsiveness. It is a concept that harnesses organisational resources to effectively manage and respond to the commotion of a volatile business environment. Agility achievement provides the means to being able to align production with demand and ultimately provide customer-driven processes. Agility concerns the rapid development of new products (or improvements to existing products) and the fast production and delivery of products to customers in response to changes in customer demand. An agile business is necessarily fast-moving and is capable of rapid response to unexpected requests and events, new opportunities and changes to customer demand requirements. Agility is essential in order for companies operating in supply chains to be able to take appropriate advantage of glass pipelines. Flexibility is the antecedent of agility (Swafford et al. 2006) as the capability for any organisational entity to be agile is founded on flexible processes. Similarly, agility is the antecedent of the customer-driven concept as the capability to deliver individualised products in high volume is founded on an agile, synchronised supply chain.

A process view of a supply chain articulates a supply chain structure in terms of three stages—procurement, manufacturing and distribution, and together the combined effect of the flexibilities of these processes repeated throughout the chain defines supply chain agility (Swafford et al. 2006). Procurement flexibility is dependent on effective supplier relationship management and is the upstream link in the iterative procurement-manufacturing-distribution sequence. The lead time of purchased items from a supplier is critical to support the agile capabilities of a downstream manufacturing partner. Manufacturing can be regarded as being agile if its modus operandi efficiently changes in response to uncertain and changing demands placed upon it (Narasimhan et al. 2006). The alignment of production with demand concepts such as the minimisation or elimination of changeover times, sequenced production, pull systems, single-piece flow and cellular manufacturing, and new developments in rapid prototyping such as polymer powder printing (Wagner 2008) are helping support the responsiveness of manufacturing systems and processes. Distribution is the downstream link in the procurement-manufacturing-distribution sequence and flexibility in this function concerns the ability to adapt the process of controlling the flow and storage of materials, finished goods and services from origin to destination in response to changing market conditions (Swafford et al. 2006).

Agility is regarded as being most desirable downstream of the position of the customer-order decoupling point in a supply chain. This allows product customisation to be postponed until late in the product build process which mitigates risk by balancing and providing short lead times with relatively high product customisation. Effective management of the customer-order decoupling point is an important factor in fostering agility. In addition, agility strategies require supply chain organisational structures to have low levels of inertia so decisions can be made quickly. Other supply chain attributes that are desirable and conducive to supply chain responsiveness include co-operative supplier relationships, processes that are flexible, short lead times and a re-configurable, dynamic approach to the management of capacity.

The inculcation of agility in a supply chain is a necessary consequence of the recognition of the need to achieve multi-tier synchronisation and/or the customisation of products. The customer-driven concept demands that agility at the supply chain or network level appropriately involves all supply chain businesses. Therefore, expanding responsiveness beyond a single process or business unit requires the generation of business initiatives that can make feasible and sustain agility across multiple organisations. Such initiatives demand a future business landscape where supply chains are attuned with mass customisation regimes, synchronising supply with demand, routinely dealing with small-lot, high-variety production but also producing and supplying customised items in single units at low cost and high efficiency. This approach to supply chain or multi-organisational agility requires the design and development of network structures that are inertia-free and operate with collaborative end-to-end visibility of demand and capacity. The complete understanding and visibility of the supply pipeline is necessary in order to guarantee individualised, customer-centric experiences.

However, agility achievement has always been recognised as a significant challenge. In 1998, Olesen (Olesen 1998) concluded that the pathway to agility for a supply network and the change processes required have to incorporate a significant number of best business practices. In 2006, Ismail and Sharifi (Ismail and Sharifi 2006) concluded that agility in supply chains goes beyond simplistic solutions such as enhancing the capabilities of individual organisations and requires an holistic approach. Today, and in the future, increasing market turbulence is accelerating clock speed leading to an even more challenging pathway to agility that demands a co-ordinated, multi-faceted effort to support, realise and create a truly customer-driven supply chain. The related concepts of postponement and mass customisation are conspicuous components of this effort.

3.3 Postponement and Mass Customisation

3.3.1 Mass Customisation and the Customer-Order Decoupling Point

Businesses are under constant pressure to meet specific customer requirements in the shortest time and at the lowest costs possible. Achieving complete product satisfaction whilst providing a commensurate and innovative service-level experience are accepted elements of contemporary business strategies. According to

Boyer et al. (2005), the modern era is one where services and products are designed to create an experience that resonates with customers. Mass production has given way to mass customisation in order to create such customer experiences. Mass customisation has been defined as the ability to provide customised products or services through flexible processes in high volumes and at reasonably low costs (Da Silveira et al. 2001) and previously as "developing, producing, marketing, and delivering affordable goods and services with enough variety and customization that nearly everyone finds exactly what they want" (Pine 1993). The agile achievements of organisations in their operations and supply chains and the promotion of the concept have catalysed the adoption of mass customisation strategies. The relationship between supply chain design and mass customisation is pivotal to the customer-driven concept. The adoption of mass customisation initiatives can require major changes to the way supply chains are organised. In recent years, supply chain design has had to cope with numerous initiatives conceived to attempt to improve business performance but have also impacted upon the ability to mass customise. Examples include outsourcing (Bardhan et al. 2007), pull systems (Productivity Development Team 1998), modularisation (Mikkola and SkjØtt-Larsen 2004), geographically proximate production (Lyons et al. 2006), postponement (Yang and Burns 2003) and shifting the customer order decoupling point (Lyons et al. 2006).

However, the term 'mass customisation' is not new. Feitzinger and Lee (1997) identified the key to mass customisation as being the postponement of product differentiation until the last possible moment; Duffell (1999) advocated the use of sales configurators to facilitate customisation; Beatty (1996) claimed that mass customisation has only became possible with the arrival of techniques such as just-in-time and total quality management. Although mass customisation has been promoted by the academic community and a lot of research concerning the topic is fundamentally theoretical in nature, there are many industrial examples of mass customisation today (see Chap. 4).

The most significant inhibitor to the implementation of mass customisation thinking is not the inadequacies of today's processing technologies, the highest hurdle is the responsiveness of today's supply chains. Even where customisation does exist and where research has been undertaken, it is only at the more downstream tiers, that is, for example, within the original equipment manufacturer (OEM) where attempts are being made to be to be sufficiently flexible. The idea of total supply chain synchronisation, where every participant in the supply of raw materials, component parts, sub-assemblies and assemblies is connected and can respond in concert to the unique requirements of an end consumer is attainable only through embracing the concept of the glass pipeline and achieving synchronisation through agility.

The decoupling point, also known as the customer-order decoupling point (CODP) (Hoekstra and Romme 1992), the demand penetration point (DPP) (Christopher 1998) and the order penetration point (OPP) (Olhager 2003), is the position in a supply chain that separates those activities that are driven by customer orders from those that are forecast-driven (see Fig. 3.6). The decoupling

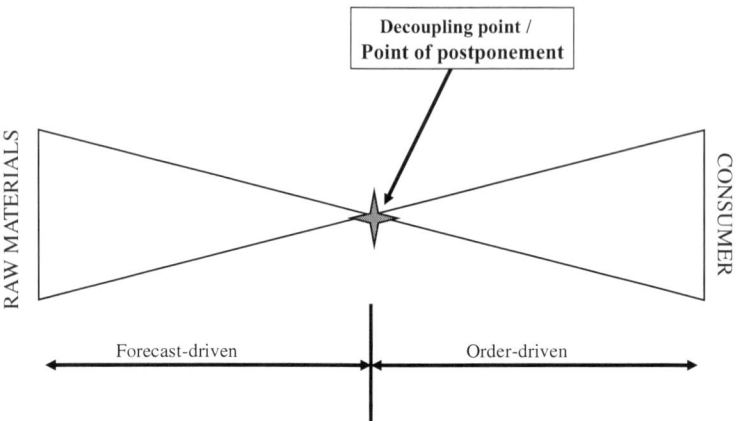

Fig. 3.6 The separation of forecast-driven and order-driven activities

point has been described as coinciding with an important inventory point from which the customer is supplied (Hoekstra and Romme 1992). This is a significant characteristic of the decoupling point as it facilitates postponement and, as a consequence, is a critical element of the customer-driven supply chain design. The further downstream or closer the decoupling point is positioned to the consumer (refer to Fig. 3.6) the more delayed or 'postponed' the differentiation of the product build or supply chain value stream. Similarly, the further upstream the decoupling point is positioned, the earlier the differentiation of the product build or supply chain value stream. (A value stream is that part of a supply chain that is responsible for producing and delivering a specific product or product family.) Early, upstream decoupling points are order biased; late, downstream decoupling points are forecast biased. Postponement is a key antecedent to mass customisation as the position of the decoupling point dictates the degree of customisation offered. Postponement can also makes use of product modularity so that a series of generic, standardised or interchangeable units can be produced to a point within the chain. Such an approach reduces the risk of inventory obsolescence.

The position of the decoupling point is crucial to the management of risk and the costs associated with providing differentiated products. It depends on "a balance between the product type, market, process and stock characteristics" (Yang and Burns 2003). Kundu et al. (2008) regard product structure, product differentiation, product variety, product postponement, component and part commonality, and component and part standardisation as influencing the positioning of decoupling points. The expectation is that the lead time from the decoupling point to the consumer is less than the market delivery time. Make-to-order (MTO), assemble-to-order (ATO) and make-to-stock (MTS) supply chain approaches result in different decoupling point positions on the raw material to consumer supply chain continuum. However, it should be noted and has been reported (Olhager 2003) that most work concerning decoupling points has been concerned with manufacturing operations rather than supply chains.

3.3 Postponement and Mass Customisation

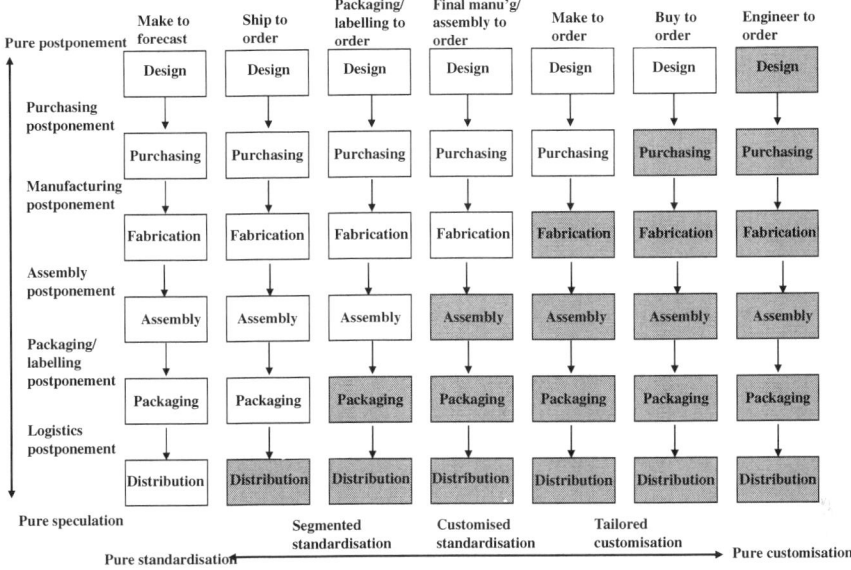

Fig. 3.7 Decoupling point positions (adapted from Yang and Burns (2003))

Figure 3.7 is adapted from Yang and Burns (2003) (in turn adapted from Lampel and Mintzberg (1996). It displays a recursive continuum of activities that combines and connects speculation and postponement with customisation and standardisation. The boundary between the shaded and unshaded portions of the framework represents the position of the decoupling point in each continuum. The further upstream the decoupling point is positioned the higher the proportion of order-driven activities.

Lampel and Mintzberg (1996) extended the three classical decoupling point positions (MTO, ATO and MTS) to include four further positions (see Fig. 3.7). Yang et al. (2007) explored these different positions in order to examine decoupling point positions and postponement strategies from an inter-organisational perspective. The authors also emphasised the implications of postponement strategies on outsourcing operations and creating alliances that produce supply networks, and highlighted the implications of moving decoupling points upstream and downstream. The position of a decoupling point within a supply network and the motivation for its movement upstream or downstream and the subsequent impact such movement has on the notion of mass customisation is inextricably linked to the customer-driven potential of the supply chain.

Rudberg and Wikner (2004) provided an analysis of the role of the decoupling point within a mass customisation context. They noted the further downstream the decoupling point is positioned the higher the degree of emphasis on productivity in operations implying price (cost) to be an expected key competitive priority; the further upstream the decoupling point the higher the emphasis on flexibility.

"Introducing a mass customisation strategy requires a technological change that moves the equilibrium point further upstream to provide a higher degree of flexibility, while at the same time increasing productivity" (Rudberg and Wikner 2004). The researchers highlighted the production and engineering dimensions in the analysis of decoupling points and the feasibility of customer involvement in both dimensions. Their research produced a two dimensional representation of a "decoupling point space" to articulate the level of customer involvement in the value chain and as a way to combine both flexibility and productivity features.

However, meeting customers' needs now involves all parties of the supply chain. Worldwide, several organisations have been successful at developing a competitive edge that relies on innovative supply chain design. Examples include organisations such as Zara and Hewlett-Packard. Success from a supply chain perspective requires that all tiers embrace a series of principles that can guarantee a fast response to customers' needs and that necessarily implies reducing the total lead time from the moment a customer places an order to the moment it is received. In addition, the supply chain must be robust and efficient enough to ensure a problem-free material flow. This implies a supply chain that is capable of reducing waste along the material flow path to minimum levels and which is responsive and agile to customer needs. Figure 3.8 is structurally identical to Fig. 3.6 but incorporates Fisher's efficient/responsive paradigm (see Chap. 1).

Upstream of the decoupling point (refer to Fig. 3.8) is the efficient, lean or 'push' portion of the chain, downstream is the responsive, agile or 'pull' portion. "The lean-agile approach combines the best from both of these paradigms: bringing together the physical efficiencies of the lean philosophy with the flexibility and customer responsiveness of the agile philosophy" (Kundu et al. 2008). Such an arrangement has been referred to as 'leagile' (Naylor et al. 1999) and is analogous to combining customer-driven processes with the achievement of agility. Mass customisation requires the supply chain to be responsive downstream of the decoupling point in order to quickly respond to customer demands but, at the same time, to be efficient and cost effective upstream of the decoupling point. Benetton is a much-cited example of a leagile supply chain, where a plain garment is the generic product from which the final coloured garment is produced. Stock of the plain garment sits at the decoupling point beyond which the final product can be produced to customer demand. In such a manner, the notion of the elimination of waste can sit comfortably alongside the notion of aligning production with demand. The critical question concerns the location of the decoupling point and, therefore, the most suitable position for the efficiency/responsiveness boundary.

Olhager (2003) highlighted the use of the ratio of production lead time to delivery lead time (P/D ratio) and the relative demand volatility (RDV) as inputs to the strategic placement of the decoupling point. The author also identified three different categories of factors that influenced the decoupling point placement The first category concerns market-related factors such as delivery lead-time requirements, product demand volatility, product volume, product range and product customisation requirements, and customer order size and frequency. The second category concerns product-related factors such as modular product design and

3.3 Postponement and Mass Customisation

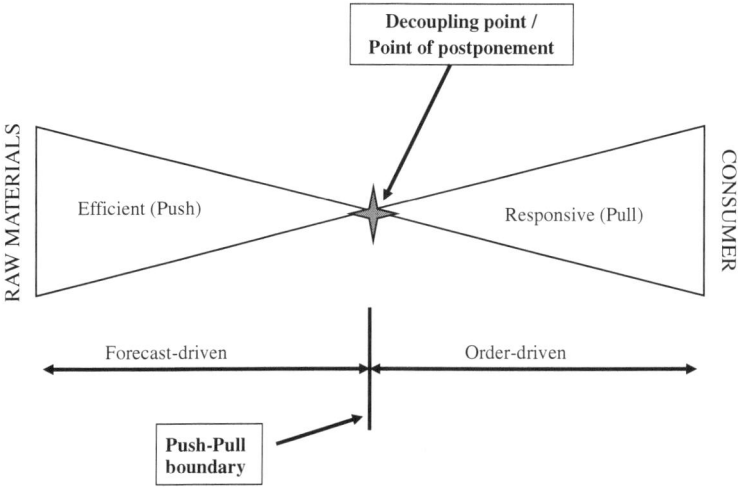

Fig. 3.8 The separation of efficient and responsive activities

customisation opportunities. The third category concerns production related factors such as production lead time, the number of planning points in a manufacturing process, the flexibility of the production process, the position of the production bottleneck and the number of resources with sequence-dependent forces. The author explained how each factor affects the position of the decoupling point and the interrelationships between them, and emphasised that any shift in the position of the decoupling point should be done in order to support a competitive priority.

Most researchers discuss decoupling points within the context of a linear operation or supply chain but Sun et al. 2008 used multiple decoupling points to partition a supply network and developed a mathematical model in order to identify the most appropriate positions for the decoupling points in such a network. The model is based on positioning decoupling points subject to a trade-off between maximising manufacturing efficiency upstream of the decoupling points and minimising inventory investment downstream of the decoupling points "while maintaining a high and consistent level of customer service". Coronado Mondragon and Lyons (2008) analysed the implications innovative information systems and geographical proximity of members of a supply network can have on decoupling point position to support supply chain design in a high-variety environment.

Kundu et al. (2008) used a knowledge-based approach to support the selection of decoupling points. They developed a knowledge model in the form of a network of production rules and applied it to a series of cases. The model focuses on product structure, product variety, product differentiation and postponement, component and part commonality, and component and part standardisation. In particular, decoupling point position is affected by the degree of modularity, and part standardisation and commonality in the product structure. Modularity, standardisation and part commonality influence the number of opportunities to postpone customisation until the most downstream stages in the supply chain. It was noted by the authors that this may extend beyond the manufacturing system to the distribution system.

The location of the decoupling point is a critical decision. It separates push from pull activities, controls postponement and differentiation, and lies at the point of customisation. Decoupling point position is influenced by a mix of market, product and production related factors. Amongst these factors, two issues are significant: the production to delivery lead time ratio (P/D ratio) and the relative demand volatility (RDV) (the standard deviation of demand relative to average demand (Olhager 2003)). In addition, a considerable body of work in the area is concerned with decoupling points in manufacturing without giving appropriate recognition to supply chains. The customer-driven supply chain is a strategic proposition that effectively harmonises pre and post-decoupling point operations across different supply networks under different competitive regimes.

3.4 Supply Chain Information and Communication Technologies

The provision of supply chain information transparency and tight inter-organisational collaboration through the creation of glass pipelines can be achieved through a range of different information and communication technologies. These include electronic data interchange (EDI), enterprise resource planning (ERP) systems, web-enabled systems and radio frequency identification (RFID) systems.

3.4.1 EDI

EDI can be defined as the electronic exchange of standardised business documents over a communications network that links computer systems of various trading partners (Down 2002). It has been a resolute approach to supply chain information exchange for many years and has been widely used in many different industries. However, it is its maturity and its widespread use that has led to the creation of multiple standards. The American National Standards Institute (ANSI) first introduced the ANSI X12 standard, which was augmented by several further versions for different types of industry. The United Nations created the Administration, Commerce and Transport (EDIFACT) standard, with the ambition to provide a single international language. Both ANSI and EDIFACT have evolved resulting in multiple versions which has, to some extent, undermined the reputation of EDI. Data transaction formats have also proliferated and EDI can be costly

3.4 Supply Chain Information and Communication Technologies

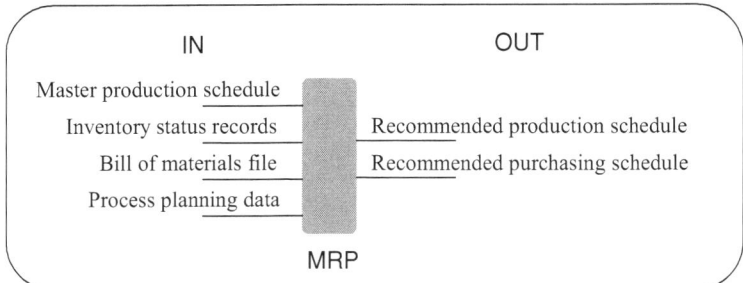

Fig. 3.9 MRP structure

to implement. However, EDI has sustained, it has proven itself to be a robust supply chain technology for quickly and metronomically exchanging data between supply chain partners and it is evolving and being further developed through the use of Internet EDI.

3.4.2 ERP Systems

An ERP system is a suite of integrated business software modules. Material requirements planning (MRP) is often regarded as the nucleus or the core module of the ERP-system suite. MRP was developed as a mechanism for companies to control the quantities of different materials purchased and different items produced for assembly operations. The MRP module in an ERP system explodes the product bill of materials (BOM) through a master production schedule (MPS), nets off on-hand inventory and takes account of production and supply lead times in order to produce recommended, time-phased production and purchasing schedules. Figure 3.9 depicts the typical inputs and outputs of an MRP system.

The BOM provides the parts' list (or recipe) and parts' structure for a particular product. The BOM composition is associated with different levels of assembly: level 0 represents the finished product, level 1 comprises the parts and sub-assemblies that constitute the finished product, level 2 comprises the parts and sub-assemblies that constitute level 1 and so on. The system logic checks the quantity of inventory at each level starting at level 0 and calculates the number of parts and sub-assemblies needed to supplement current inventory levels to meet customer demand.

The core of the MRP module is the 'MRP netting process', an algorithmic procedure that uses planning information to calculate part quantity and timing requirements. The process takes place via the BOM explosion in order to establish the number of sub-assemblies and parts required. The number of required parts available in stock is checked before moving down the BOM to the next level. The MRP module has the capability of calculating the quantities of materials required using back-scheduling which is done by taking into account the appropriate lead times associated with every BOM level. The outputs from the MRP process are the:

- 'recommended production schedule'—a detailed schedule showing the required production start and completion dates and quantities required to meet the demand from the MPS;
- 'recommended purchasing schedule'—a detailed schedule displaying the dates that purchased items should be available and the dates the ordering process should take place.

When MRP was introduced, it represented a major advance for manufacturers. However, standalone MRP systems had several shortcomings. One of them concerned data integrity. Inaccurate inventory, BOM or MPS data will generate an incorrect output, potentially affecting both the recommended production schedule and the recommended purchasing schedule. Another issue with MRP systems concerned the lead time to produce a product from its component parts. MRP assumes a constant manufacturing lead time which is not always practical. The MRP process did not always give adequate consideration to available capacity, therefore undermining the integrity of recommended schedules. The shortcomings of MRP were addressed by the development of Manufacturing Resource Planning (MRPII) systems. MRPII systems are regarded as an extension of MRP systems and typically incorporate operational and financial planning capabilities plus simulation capabilities that provide the means to scenario-test.

ERP systems take advantage of their capabilities to automate interactions in order to provide a common source of data for cross-functional connectivity and co-operation. SAP, Oracle and Peoplesoft are some of the best-known ERP system providers. Key connectivity characteristics of ERP systems include:

- compatibility with most common platforms including UNIX, Linux and Windows;
- interfaces to common application programmes such as spreadsheets so data can be conveniently imported and exported;
- web compatibility;
- links to information transmission technologies such as EDI.

ERP systems can not only connect different functions and activities, such as MRP, within a business entity but also can be used to share information with external partners such as suppliers, distributors and third-party logistics providers. In addition, most ERP systems comprise human resource management, quality management, product engineering, maintenance management, costing, advanced planning and scheduling, just-in-time support, logistics and distribution, demand management, decision support, sales planning and financial planning functions. In recent years, advanced ERP systems have incorporated additional functions such as customer relationship management (CRM), supplier relationship management

3.4 Supply Chain Information and Communication Technologies

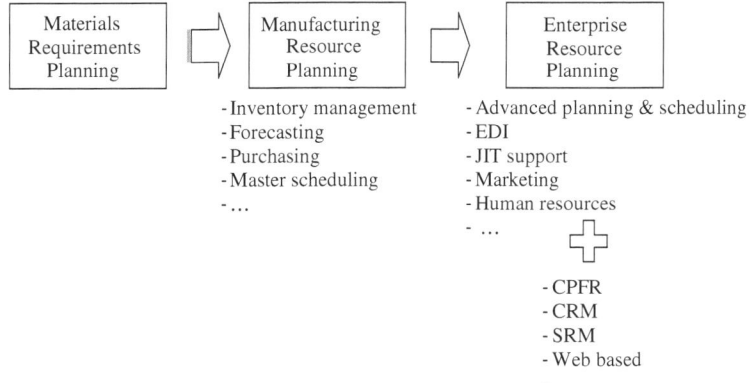

Fig. 3.10 Evolution from MRP to extended ERP

(SRM), collaborative planning, forecasting and replenishment (CPFR), automatic identification and data capture (AIDC) and web-based support through portals, EDI and XML. The evolution from MRP-based systems and the growing sophistication of ERP systems is represented by Fig. 3.10.

The added functionality found in ERP systems such as CRM or SRM has created new opportunities for supply chain integration. For example, the CRM module of an ERP system can be used to relationship build with customers, assess their potential profitability, and determine the products and services necessary to maintain a loyal customer base (APICS 2006). SRM facilitates the supply needs of an organisation and an SRM module can assist in fostering close relationships with key suppliers. This may require the marketing department to book orders, the engineering department to define materials to be used, the production department to define manufacturing processes and the logistics department to develop a plan to move materials efficiently (APICS 2006).

According to the APICS Certified Supply Chain Professional Program (APICS 2006) the added functionality of ERP systems provides the opportunity for:

- "better business management" through analytical tools for the manipulation and transformation of data;
- "performance measurement" through facilities for the definition, measurement and control of performance;
- "addition of operational initiatives" through the support of competitive paradigms such as build-to-order (BTO) and lean manufacturing methods;
- "industry-wide and industry-specific solutions" through the use of modules that are generic such as supply chain planning and those that are sector specific such as VMI for retailers;
- "on-demand access to data" through the use of distributed architectures;
- "improved effectiveness" through process automation;
- "value-chain opportunities" provide external collaboration through the provision of CRM, SRM, portals, XML and EDI.

Despite the difficulties many organisations have encountered with the implementation of ERP systems, the truth is that for many of them operating without an ERP system would be very difficult. ERP has enabled numerous organisations to have enhanced connectivity with customers and suppliers. Also ERP has promoted the standardisation of business processes and subsequently contributed significantly to improved business process efficiency.

3.4.3 Web-Enabled Systems

The Internet is an important supply chain umbilical providing large-scale connectivity between supply chain partners. Internet applications linking different companies provide obvious glass pipeline opportunities and have made it possible to overcome some of the difficulties associated with the adoption of technologies such as EDI. The high investment costs associated with the introduction of EDI have often precluded its adoption in many small and medium-sized enterprises (SMEs). However, Internet EDI transmissions can potentially provide a productive combination of web browser capability and structured document formats, and yet cost significantly less than conventional EDI transmissions.

The two principal methods of undertaking Internet-facilitated EDI are 'Internet EDI' and 'WEBEDI'. Internet EDI utilises the Internet as a platform for transactions with EDI data integrated into internal information systems. E-mail services or file transfer protocol technologies (FTP) are used for this type of EDI (Lyons et al. 2005). WEB EDI makes use of the World Wide Web to create an electronically-linked community. WEBEDI initiatives do not make use of EDI standards, rather via technologies such as EDI translators and web browsers, data can be transformed into EDI transactions.

Internet applications and e-commerce have benefited directly from developments in 'middleware technology' and supply chains have made use of middleware and distributed computing through the implementation of information architectures that are robust and provide integration opportunities. For example, XML/Java applications are now mature tools that are used to insert and extract data into and from web servers. With reference to ERP systems and the Internet, organisations are aware of the importance for ERP solutions to provide full support to Internet-based communication channels. Having presence on the Internet has become an integral part of doing business for almost all organisations. E-business does not solely concern the provision of a sales channel but has a significant use as a means to foster supply chain collaboration between suppliers, manufacturers and customers.

E-business can be regarded as a group of Internet-enabled practices that manage the flow of materials, information and credit along the different tiers that comprise the supply chain. E-commerce can be regarded as a sub-set of e-business concerned principally with the tasks of buying and selling goods and/or services.

3.4 Supply Chain Information and Communication Technologies

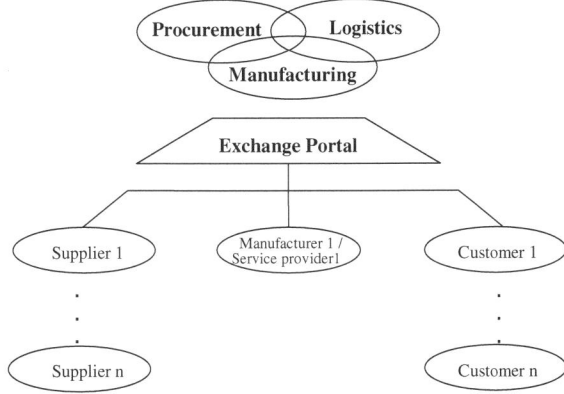

Fig. 3.11 Exchange portal and the supply chain

Original e-commerce applications started as static websites but have evolved to such an extent that contemporary applications are fully interactive, where orders can be placed, goods purchased, paid for and tracked on-line. Also, e-commerce has re-shaped for good the structure of many supply chains, deleting whole processes and entire tiers. A simple example concerns airlines selling flight tickets on the Internet directly to the end customer, eliminating the role of intermediaries such as travel agents.

As e-commerce has matured, there has been a proliferation of portals and on-demand services. Portals are integral to e-business models and provide access to an organisation's systems via web browsers. Increasing Internet access has also enabled the development of the exchange portal, a communication medium that has important implications for supply chain communication (Bowersox et al. 2002). One of the key characteristics of an exchange portal is that it facilitates horizontal and vertical information exchange between supply chain partners. This information exchange may involve manufacturing, procurement and logistics functions. Figure 3.11 depicts the use of an exchange portal designed to facilitate communication between companies in a supply chain. In the figure, an indefinite number of suppliers can participate. The same applies to the customers of the manufacturer/service provider. Portal solutions can be found in many sectors including aerospace, food, construction, retail and automotive.

On-line marketplaces are also part of the modern-day e-business portfolio. These sites are created with the specific purpose of bringing together buyers and sellers with the aim of creating a large market with low transaction costs. However, the rationale for establishing marketplaces has shifted from extending procurement to a much broader range of opportunities for supply chain collaboration and synchronisation (Berger and Gattorna 2001). E-marketplaces, or on-line marketplaces, can act as a fast-track mechanism for creating effective inter-organisational collaboration and integration.

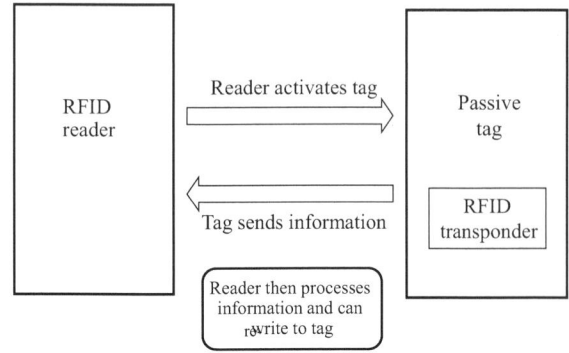

Fig. 3.12 Illustration of a passive RFID tag arrangement

3.4.4 RFID

The sophistication of information technology present in contemporary supply chain structures goes hand-in-hand with the implementation of different hardware devices. In the supply chain, RFID tags are used in pallets and crates providing traceability from a point of origin through the logistics system and into retail stores. RFID tags are classified as either active or passive. Active tags are similar to the wireless nodes of sensor networks. Sensor-network nodes include sensors to measure and provide information on the properties (such as temperature and humidity) associated with their environments, whereas active tags focus on providing only an identification number and basic information (such as a description or transportation history) relating to the tagged object.

The key characteristic of a passive tag is that it provides only an identification number when an RFID reader 'beams' the tag with radio waves. The beam from the reader activates the tag whilst operating as a communication link between the RFID reader and the tag (Borriello 2005). Figure 3.12 depicts the interaction between a passive tag and an RFID reader. RFID tags are becoming cost-effective enough to compete with printed bar codes (Borriello 2005).

3.4.5 Other Supply Chain Information Systems

Mobile technologies are being increasingly integrated into supply chain information systems. Global Positioning Systems (GPS), General Packet Radio Services (GPRS) and Geographic Information Systems (GIS), in conjunction with Internet applications, are providing 'closed-loop' glass pipeline supply chain transparency through the communication of real-time localisation and traceability of shipments, and the status of product deliveries (Michaelides et al. 2010). GPS is conspicuous in the transport and delivery of small, individual items such as parcels where the market leaders such as DHL and UPS have used real-time item tracking

3.4 Supply Chain Information and Communication Technologies

for many years (Malladi and Agrawal 2002; Michaelides et al. 2010). Vehicle tracking is also a common application for GPS devices. Vehicle schedules and routes can be re-planned at short notice while the vehicle is in transit and planners are able to take advantage of backloading and load consolidation opportunities in real time (Waters 2007). This improves supply chain efficiency and customer service and responsiveness.

Michaelides et al. (2010) reported on a case study they had undertaken focusing on the relationship between a road transport haulier and a transatlantic shipping company. The haulier had identified persistent problems caused by the inability to effectively track containers, and noted that improving the accuracy and availability of time estimates for containers reaching their destination would permit more effective planning. The authors of the study reported that both companies recognised the need for a simple and efficient system to track containers as a means to improve performance. The case focused on the development of an integrated GPS/portal solution. Tracking was achieved through the use of GPS-enabled handsets that were used to transmit the co-ordinates of the vehicle to the portal (Michaelides et al. 2010). The location of the vehicle was made visible to system users on a map who could then establish the estimated time of arrival at the delivery location (Michaelides et al. 2010). Proof of delivery (PoD) information was captured on a PDA and transmitted to the portal. Significantly, the solution is compatible with the glass pipeline concept as the information is available on-line across the whole supply chain and provides retrieval of previous transaction data. Furthermore, the system's interoperability capabilities facilitate exchange of information with different back-office systems resulting in a more holistic and integrated solution providing order and payment functions. Figure 3.13 depicts the solution arrangement.

'Advanced planning system' (APS) is an umbrella term for applications that provides decision support by using operational data to analyse and optimise the flows through the supply chain. Techniques utilised by an APS include forecasting and time series analysis, optimisation techniques (linear programming, mixed integer programming, location-allocation techniques, and genetic and rule-based algorithms), and scenario planning (what-if analysis and simulations).

Sometimes provided as an ERP module, inventory management systems (IMS) are used to monitor, manage and optimise inventory levels along the supply chain by controlling and adjusting safety stocks in order to provide customers with a high level of service while determining the locations where optimal levels of inventory will be held. Warehouse management systems (WMS) are used to manage material flow in and out of storage locations based on the use of pre-determined decision criteria. WMS can be seen in the use of cross-docking operations at some retailers' distribution centres, where products handled spend a short time in the storage locations.

A common practice is to customise an ERP system based on specific business needs. It is also common to find that major industry players such as aerospace original equipment manufacturers (OEMs), automotive OEMs and large retailers have developed their own proprietary systems which can be used to transmit requirements to first-tier suppliers. Proprietary applications may be used to

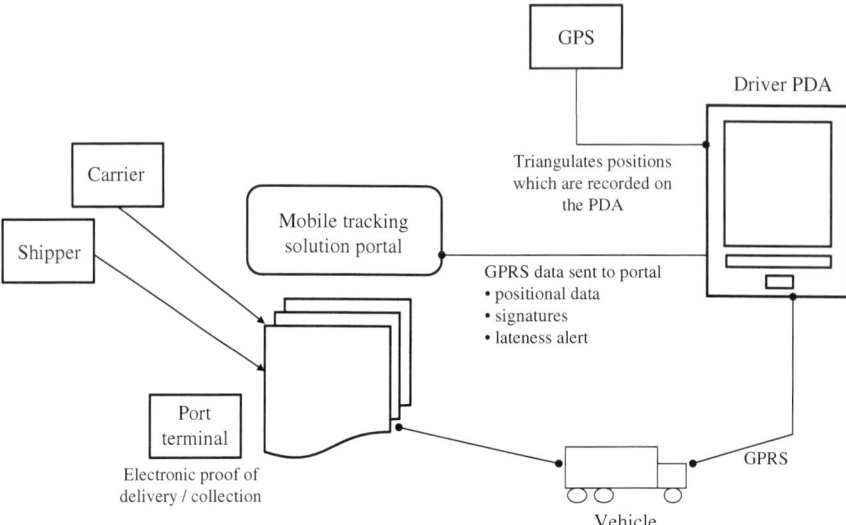

Fig. 3.13 Real-time tracking management solution (adapted from Michaelides et al. (2008), and Michaelides et al. (2010))

communicate aggregated daily requirements to suppliers. The systems provide the fixed requirements for the coming days and tentative requirements for the coming weeks and months. Also proprietary applications are used to communicate the sequencing of components, an initiative found in the automotive sector that has enabled vehicle assemblers to support synchronisation through the adoption of sequenced supply with first-tier suppliers usually located on supplier parks adjacent to the vehicle assembler's site. Figure 3.14 depicts a sequenced-supply process arrangement.

The example in Fig. 3.14 illustrates the case of a vehicle assembler and a first-tier supplier. Both partners have in place an application that enables the sequenced delivery of modules or components. For every vehicle built there is a unique set of modules which are delivered in sequence to the point-of-fit at the vehicle assembly line. At that point, a module in a dark colour combination will be installed in a vehicle with a saloon body configuration, the next module sequenced which is in a light colour combination may be installed in a vehicle with a coupe body configuration and so on.

Recent years have also witnessed an increase in the use of agent technologies affecting manufacturing and supply. A software agent is an autonomous problem-solving unit that may work with other agents to achieve optimised results in a specific problem area. Agent technologies can be used in an environment where a manufacturer and its suppliers require the use of a proprietary interface for exporting and importing data in a non-standard format. Moreover, agents have played a key role in the integration of networking technologies developed on open architectures, facilitating the automation of large-scale distributed systems.

3.5 Chapter Summary

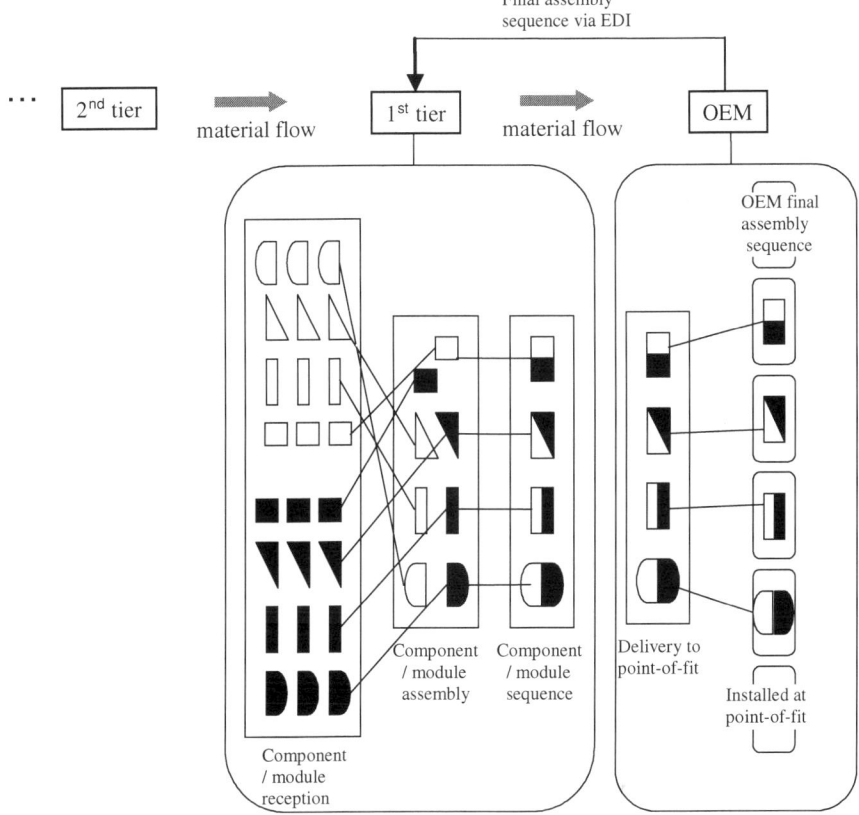

Fig. 3.14 An illustration of sequenced supply

3.5 Chapter Summary

Customer-driven supply chain management requires effective customer-driven processes and the synchronisation of planning, production and delivery activities between supply chain tiers. Supply chain synchronisation is achieved through information sharing and the agility of supply chain processes. In most supply chains, information is communicated only between contiguous and immediate tiers of suppliers in a time-phased, sequential manner. The authors have used the term 'glass pipeline' to reflect the need for information systems to support the customer-driven concept by having a commitment to sharing primary demand data and information across supply-chain tiers so that customer and supply chain behaviour is transparent to decision makers. The provision of supply chain information transparency and tight inter-organisational collaboration through the creation of glass pipelines can be achieved through a range of different information and communication technologies. These include EDI, ERP systems, web-enabled systems, RFID and mobile technologies.

Data flowing upstream in a supply chain is often unreliable because of the distortive effects of order rather than demand data being processed in multiple material requirements planning systems. This fragmented rather than integrated approach to supply chain planning inhibits effective co-ordination and results in demand amplification, inventory stockpiles, stock outs and poor customer service. The transmission of POS data from retailers to upstream tiers accompanied by a system of multi-tier VMI so that replenishment decisions are assigned to a single point between any two consecutive tiers improves co-ordination between tiers and the synchronicity of planning, production and delivery activities. This combined approach to the sharing of data and utilising VMI is an example of the innovative use of both information and material flow systems that is often imperative in achieving a customer-driven supply chain.

Effective execution of the glass pipeline concept is not only dependent on the judicious use of information and communications technology but also requires inter-organisational collaboration and trust, porous organisational boundaries and customer-driven, agile processes that can take advantage of real-time access to information. Agility is predicated on responsiveness and is essential in order for supply chain companies to be able to take appropriate advantage of glass pipelines. Agile processes facilitate rapid adaptation in response to change and so provide the means to align production and supply chain activity with demand. Effective management of the customer-order decoupling point is an important factor in fostering agility. The customer-driven concept demands that agility at the supply chain level appropriately involves all supply chain members businesses but agility is regarded as being most critical downstream of the position of the customer-order decoupling point in a supply chain. This allows product customisation to be postponed until late in the product build process which reduces risk by combining short lead times with relatively high product customisation. This supports mass customisation—the production of customer-specified products in high volume. In addition, agility strategies require supply chain organisational structures to have low levels of inertia so decisions can be made quickly.

References

APICS (2006) Using information technology to enable supply chain management, Module 4 certified supply chain professional program

Bardhan I, Mithas S, Lin S (2007) Performance impacts of strategy, information technology applications, and business process outsourcing in US manufacturing plants. Prod Oper Manag 16(6):747–762

Beatty RT (1996) Mass customisation. Manuf Eng 75(5):217–220

Berger AJ, Gattorna JL (2001) Supply chain cyber mastery. Gower Publishers, Hampshire

Borriello G (2005) RFID: tagging the world. Commun ACM 48(9):35–37

Bowersox D, Closs D, Bixby Cooper M (2002) Supply chain logistics management. McGraw Hill, New York

Boyer KK, Frohlich MT, Hult GTM (2005) Extending the supply chain: how cutting-edge companies bridge the critical last mile into customers' homes. Amacom, New York

References

Chen F, Drezner Z, Ryan JK, Simchi-Levi D (2000) Quantifying the bullwhip effect in a simple supply chain: the impact of forecasting, lead times, and information. Manag Sci 46(3):436–443

Coronado Mondragon AE, Lyons AC (2008) Investigating the implications of extending synchronized sequencing in automotive supply chains: the case of suppliers in the European automotive sector. Int J Prod Res 46(11):2867–2888

Christopher M (1998) Logistics and supply chain management strategies for reducing cost and improving service. Financial Times Prentice Hall, London

Croson R, Donohue K (2005) Upstream versus downstream information and its impact on the bullwhip effect. Syst Dyn Rev 21(3):249–260

Da Silveira G, Borenstein D, Fogliatto F (2001) Mass customisation: literature review and research directions. Int J Prod Econ 72:1–13

Dejonckheere J, Disney SM, Lambrecht MR, Towill DR (2004) The impact of information enrichment on the bullwhip effect in supply chains: A control engineering perspective. Eur J Oper Res 153(3):727–750

Disney SM, Towill DR (2003) The effect of vendor managed inventory (VMI) dynamics on the bullwhip effect in supply chains. Int J Prod Econ 85(2):199–215

Disney SM, Naim NM, Potter A (2004) Assessing the impact of e-business on supply chain dynamics. Int J Prod Econ 89(2):109–118

Down A (2002) Integrating EDI systems across and beyond your enterprise. IBM software group http://download.boulder.ibm.com/ibmdl/pub/software/cn/websphere/products//download/whitepapers/edi_v1c.pdf

Duffell J (1999) Mass customisation across the business: configurators and the internet part 2. Control 24(10):14–16

Feitzinger E, Lee HL (1997) Mass customization at Hewlett-Packard: the power of postponement. Harvard Business Review January–February: pp 116–121

Forrester J (1958) Industrial dynamics: a major breakthrough for decision makers. Harvard Business Rev 36(4):37–66

Forrester JW (1961) Industrial dynamics. MIT Press, Cambridge

Hoekstra S, Romme J (1992) Integrated logistics structures: developing customer oriented goods flow. McGraw-Hill, London

Ismail HS, Sharifi H (2006) A balanced approach to building agile supply chains. Int J Phys Distrib Logist Manag 36:431–444

Kaipia R, Holmström J, Tanskanen K (2002) VMI: what are you losing if you let your customer place orders? Prod Plan Control 13(1):17–25

Kundu S, McKay A, de Pennington A (2008) Selection of decoupling points in supply chains sing a knowledge-based approach. Proc Inst Mech Eng Part B J Eng Manuf 222:1529–1549

Lampel J, Mintzberg H (1996) Customizing cusotmization. Sloan Manage Rev 37:21–30

Lee HL, Padmanabhan V, Whang S (1997) The bullwhip effect in supply chains. Sloan Manage Rev 38(3):93–102

Lyons AC, Coronado-Mondragon AE, Bremang A, Kehoe DF, Coleman J (2005) Prototyping an information system's requirements architecture for customer-driven, supply-chain operations. Int J Prod Res 43(20):4289–4320

Lyons A, Coronado A, Michaelides Z (2006) The relationship between proximate supply and build-to-order capability. Indus Manage Data Syst 106(8):1095–1111

Malladi R, Agrawal DP (2002) Current and future applications of mobile and wireless networks. Communications of the ACM 45(10):144–146

Michaelides R, Liu K, Jervis S (2008) E-solutions enabling growth in the transport industry: case study of a real- time tracking management system. Proceedings of the thirteenth LRN conference, Liverpool, pp 291–296

Michaelides R, Michaelides Z, Nicolaou D (2010) Optimisation of logistics operations using GPS technology solutions: a case study. Proceedings of the twenty first annual POMS conference, Paper number 015-0822, May, Vancouver, Canada

Mikkola J, Skjøtt-Larsen T (2004) Supply-chain integration: implications for mass customization, modularization and postponement strategies. Prod Plan Control 15(4):352–361

Moyaux T, Chaib-draa B, D'Amours S (2007) Information sharing as a coordination mechanism for reducing the bullwhip effect in a supply chain. IEEE Trans Syst Man Cybernet Part C Appl Rev 37(3):396–409

Narasimhan R, Swink M, Kim SW (2006) Disentangling leanness and agility: an empirical investigation. J Oper Manage 24:440–457

Naylor JB, Naim M, Berry D (1999) Leagility: Integration the lean and agile manufacturing paradigms in the total supply chain. Int J Prod Econ 62:107–108

Olesen JD (1998) Pathways to agility: mass customization in action. John Wiley, New York

Olhager J (2003) Strategic positioning of the order penetration point. Int J Prod Econ 85(3):319–329

Patton S (2005) Supply chain the perfect order. CIO Magazine, August 01 2005

Pine BJ (1993) Mass customization: the New Frontier in business competition. Harvard Business School Press, Boston

Productivity Development Team (1998) Just-in-Time for operators. Productivity Press, Portland

Rudberg M, Wikner J (2004) Mass customization in terms of the customer order decoupling point. Prod Plan Control 15(4):445–458

Schonberger RJ (2006) Supply chains: tightening the links. Manuf Eng 137(3):77

Simchi-Levi D, Kaminsky P, Simchi-Levi E (2007) Designing and managing the supply chain: concepts, strategies and case studies. McGraw Hill, New York

Souza C (2003) Cisco regulates inventory intake. http://www.eetimes.com/electronics-news/4032689/Cisco-regulates-inventory-intake. Accessed Dec 2010

Sterman J (1989) Modeling managerial behaviour: misperception of feedback in a dynamic decision-making experiment. Manag Sci 35(3):321–339

Swafford PM, Ghosh S, Murthy N (2006) The antecedents of supply chain agility of a firm: scale developments and model testing. J Oper Manag 24:170–188

Sun XY, Ji P, Sun LY, Wang YL (2008) Positioning multiple decoupling points in a supply network. Int J Prod Econ 113(2):943–956

Wagner S (2008) Power of print. The Engineer March–April, pp 32–32

Waters D (2007) Trends in the supply chain: 1–20. In: Waters D (ed) Global logistics: new directions in supply chain management. Kogan Page, London

Yang B, Burns N (2003) Implications of postponement for the supply chain. Int J Prod Res 41(9):2075–2090

Yang B, Yang Y, Wijmgaard J (2007) Postponement: an inter-organizational perspective. Int J Prod Res 45(4):971–988

Chapter 4
Mass Customisation: A Strategy for Customer-Centric Enterprises

> This chapter combines and extends the arguments developed in two earlier publications: Salvador, de Holan and Piller (2009) and Piller (2008).

4.1 Introduction

It is through leveraging the potential of the supply chain to satisfy consumers' needs that is increasingly determining business competitiveness. This notion embraces and extends the very essence of Drucker's statement that "it is the customer who determines what a business is" (Drucker 1954). In many industries, businesses are being confronted by an uninterrupted trend towards heterogeneous customer requirements and changeable market conditions. Explanations for this may be found in the growing number of single households, changing demographic structures, an orientation towards design, and a new awareness of service, quality and functionality that demands innovative and reliable products corresponding exactly to the specific customer needs (Zuboff and Maxmin 2003; Anderson 2007; Franke et al. 2009). In particular, those consumers with significant purchasing power are increasingly trying to express their personalities through the specificity of their possessions. Thus, manufacturers are being obliged to create portfolios of products with an increasing wealth of variants. As a consequence, many companies are having to manage their customers' demand on an individual basis.

To address this challenge, contemporary technologies are providing opportunities that have previously not been available. Modern information and communication technologies, for example, provide glass pipeline opportunities that have enabled pervasive connectivity and direct interaction between and amongst individual manufacturers, resellers, consumers, suppliers and logisticians. This connectivity offers enormous flexibility. Beyond "listening into the customer domain" (Dahan and Hauser 2002) in order to better address specific needs and to reduce response times, manufacturers have the means to regard customers as individuals and to proactively develop products which cater to them at the prices they are willing to pay and within the time frames that they are willing to accept. Yet despite all the technological advances, this is by no means a straightforward task. Particularly in today's highly competitive business environments, activities for serving customers have to be performed both efficiently and effectively—they have to be organised around a customer-centric supply and demand chain with glass pipeline-enabled, agile processes.

Since the early 1990 s, mass customisation has emerged as a leading idea for achieving precisely this objective and, as has already been highlighted in Chap. 3, it is a key component of the effort to achieve a customer-driven supply chain. In accordance with Pine (1993) we define mass customisation as "developing, producing, marketing, and delivering affordable goods and services with enough variety and customization that nearly everyone finds exactly what they want." In other words, the goal is to provide customers what they want, when they want it. Hence, companies adopting mass customisation regimes are necessarily becoming customer-centric enterprises (Tseng and Piller 2003), organising all of their value creating activities around interactions with individual customers.

However, to apply this apparently simple idea in practice is quite complex. As a business paradigm, mass customisation provides an attractive proposition to add value by directly addressing customer needs and in utilising resources efficiently without incurring excessive cost. This is particularly important in circumstances where competition is no longer just based on price and conformance to quality. When the subject of mass customisation is raised, the successful business model of the computer supplier Dell is often cited as one of the most conspicuous examples. It has been a recognised mass customisation exemplar. The growth and success of Dell has been based on its ability to produce and execute its business model of providing custom computers on demand, meeting precisely the needs of each individual customer and producing these items, with no finished goods' inventory risk, only after an order has been placed (and paid for). However, beyond Dell, there are many other examples of companies that have employed mass customisation successfully. For example:

- Pandora Radio© relieves people of having to channel surf through radio stations to find the music they like. Customers submit an initial set of their preferred song choices, and from that information Pandora identifies a broader set of music that fits their preference profile and then broadcasts those songs as a custom radio channel. As of summer 2010, Pandora.com had 48 million listeners who created more than half a billion radio stations from the 700,000 tracks in its library and who listened, on average, for 11.6 h per month.
- BMW customers can use an online toolkit to design the roof of a Mini Cooper with their very own graphics or picture, which is then reproduced with an advanced digital printing system on a special foil. The toolkit has enabled BMW to tap into the custom after-sales market, which was previously owned by niche companies. In addition, Mini Cooper customers can also choose from among hundreds of options for many of the car's components, as BMW is able to manufacture all cars on-demand according to each buyer's individual order.
- Selve, a London and Munich-based supplier of custom-made, women's shoes, is a perfect example of a company that interacts well with its customers via a traditional retail sales channel and also online. Selve provides its customers with

4.1 Introduction

the means to create their own shoes by selecting from a variety of materials and designs, in addition to providing a truly custom-fit based on a 3D-scan of a customer's feet. Trained consultants provide advice in the company's stores; the online shop offers re-orders. Shoes are all made to order in a specialised factory in China and are delivered in about two weeks. Customers get this dedicated service for a cost between 150 and 450 Euros, not inexpensive but still affordable for many consumers compared to the price level of a traditional shoe maker.

- Sears© has become one of the leading players in the customisation and personalisation business in the United States. Its affiliate company Lands' End was one of the first companies to offer mass customisation of garments online. Today, up to 60% of all Lands' End products in certain categories are truly made to order. However, in its appliances business, under the Sears and Kenmore brand, the company is a forerunner in offering online toolkits with which consumers can design their own kitchens (and other rooms) and help them to match the best fitting furniture and appliances from Sears' wide range.

What do these examples have in common? Regardless of product category or industry, they are all customer-driven and have all turned customers' heterogeneous needs into an opportunity to create value, rather than regarding heterogeneity as a problem that has to be minimised, challenging the 'one size fits all' assumption of traditional mass production. This chapter explores the common elements and characteristics of successful mass customisation strategies. The core message we want to deliver here is that mass customisation should not be seen as a dedicated business model or a specific form of competitive strategy. Rather, mass customisation is the development of processes and capabilities for aligning an organisation and its supply chain with its customers' needs. We will introduce three fundamental capabilities which characterise mass customisation (Salvador et al. 2009):

- the ability of an organisation to identify the product or service attributes along which customer needs diverge ('solution space definition'),
- the ability to reuse or recombine existing organisational and supply chain resources ('robust process design'),
- the ability to help customers identify or build solutions to their own needs ('choice navigation').

The remaining sections of this chapter are organised as follows. In the next section, we review the development of the notion of customer centricity. We then explore the term mass customisation and comment on its development in the literature. The three underlying capabilities of mass customisation are then discussed in more detail. Here, we also discuss specific methods to implement these capabilities. The chapter closes with some remarks on the implementation of mass customisation in industrial practice.

4.2 The Development of Customer Orientation and Customer Centricity

The idea of a customer-centric enterprise and customer-centric supply chain is to focus all business unit and supply chain operations on serving customers and delivering unique value by treating customers as individuals (Sheth et al. 2000; Tseng and Piller 2003; Piller et al. 2006). To offer a better understanding of the specifics of customer centricity, we will briefly review in this section the development of the modern business from mass production origins to recent developments in customer orientation and customer-driven supply chains.

Mass customisation is often believed to be a strategy for replacing mass production (Pine 1993). However, before mass production was brought about by the Industrial Revolution, products were being customised with 'craftsmanship'. Each customer was a segment of one, and 'marketing' was individualised and personal, but performed implicitly and as part of the interaction process. Craftsmanship often produced high quality products that were only available to select groups of individuals (that is, those with the appropriate purchasing power). The advent of 'mass production' standardised the products and operations to leverage economies of scale and the division of labour. This dramatically reduced the cost of production. As a consequence, a large part of the population could now afford goods and services that had only been available to a small section of society. A new generation of mass consumers grew up to enjoy the products that were designed to meet the demands of a segment of the population large enough to justify the fixed costs of production, including set-up costs and capital outlays. The "mass consumption society" (Sheth et al. 2000) developed itself into a seller's market, leading firms to adopt organisational structures centred on products. Groups of related products were regarded during this period as the primary basis for structuring an organisation (Homburg et al. 2000).

The resulting increase in product variety and increasing competition by the end of the 1950 s led businesses to start paying more attention to markets than to products. 'Market orientation' emerged as an organisational pattern for firms, following Drucker's (1954) argument that creating a satisfied customer is the only valid definition of business purpose. The first objective of market orientation is to uncover and satisfy customer needs at a profit. Kotler (1991 [1967]) popularised the market-oriented perspective, and it soon became widely adopted. Market orientation implies a perception of the total market not as a homogeneous mass market, but as a collection of market segments of consumers. 'Segmentation' started with the notion of socio-demographic division with variables such as age, sex, and income. This resulted in a limited number of focused product variants (Smith 1956). Later, segmentation became more refined. More subtly defined niches based on lifestyles and previous buying behavior resulted in an increasing number of product variants for catering to specific needs. Market segmentation demands information on consumers' needs (Narver and Slater 1990). Today's instruments of market research were created as tools to satisfy precisely this set of

4.2 The Development of Customer Orientation and Customer Centricity

demands by securing pertinent information about customers for better understanding their needs.

Continuous refinement of the segmentation concept led to market segmentation being replaced by the notion of 'customer orientation'. The principal features of this strategy are (i) a set of beliefs that puts the customer's interests first; (ii) the ability of the organisation to generate, disseminate, and use superior information about customers and competitors; and (iii) the co-ordinated application of interfunctional resources to the creation of superior customer value (Day 1994). In particular, the strong emphasis on providing 'customer value' in all functions of an organisation can be regarded as the differentiation of customer orientation versus the previous stage of market orientation. The customer moved firmly into focus. During this time, the notion of the marketing function as the central entity to deal with customers developed. 'Relationship management' reinforced this perspective. It "emphasizes understanding and satisfying the needs, wants, and resources of individual consumers and customers rather than those of mass markets or mass segments" (Sheth et al. 2000). Rather than segments of customers, individual customers were seen as the target of the marketing mix, resulting in the term "one-to-one marketing" (McKenna 1991). The members of one market segment are now no longer regarded as homogeneous in relation to their profit contribution for the business, but each customer is assessed individually. Based on an individual output-to-input ratio of the marketing function for individual customers ('share of wallet'), customers are addressed either by a standardised offering or, if it pays off, by a customised offering (Parasuraman and Grewal 2000). As a result, product-based strategies are being replaced by a competitive strategy approach based on growing the long-term customer equity of a business.

The 'customer-centric enterprise' combines the organisational perspective of customer orientation with the individual perspective of relationship management (Tseng and Piller 2003; Piller et al. 2006). It also extends the responsibility of dealing with customers from solely the marketing function to the entire organisation. Customer centricity demands that the organisation as a whole is committed to meeting the needs of all relevant customers. We have attempted to extend this notion to the network or supply chain. At the strategic level, this translates to the orientation and mindset of a business or network of businesses towards sharing interdependencies and values with customers over the long-term. At the operational level, companies have to align their processes with the customers' requirements, instead of focusing solely on the requirements of their own internal operations. Such a requirement demands appropriate co-operating practices and systems, and an enlightened leadership style. These changes include an inertia-free, customer-centric organsation structure. Traditionally, separated functions such as sales, marketing (communications), and customer service will be integrated into one customer-centred activity (Sheth et al. 2000). Further, customer centricity is switching the marketing perspective from the demand side to the supply side (Piller et al. 2006). Marketing management has traditionally been viewed as demand management. The focus has been on the product or the market, and marketing has had to stimulate demand for an offering through promotional activities such as incentives or pricing policies. The customer-centric enterprise switches its focus to the individual customer as the starting point for all activities. Instead of creating and stabilising demand—that is, trying to influence people in terms of what to buy, when to buy, and how much to buy—businesses should try to adjust their capabilities including product design, production, sales, and supply chain design to respond to customer demand. Mass customisation can be seen as a way of thinking for companies to achieve these goals of customer centricity, both with regard to marketing and sales as well as to operations and supply chain management.

4.3 Mass Customisation: Definition and Literature Review

From a strategic management perspective, mass customisation is a differentiation strategy. Referring to Chamberlin's (1962) theory of monopolistic competition, customers gain the increment of utility of a customised good that better fits their needs than the best standardised product attainable would. The larger the heterogeneity of all customers' preferences, the larger is this gain in utility (Kaplan et al. 2007). Davis, who initially coined the term in 1987, refers to mass customisation when "the same large number of customers can be reached as in mass markets of the industrial economy, and simultaneously […] be treated individually as in the customized markets of preindustrial economies" (Davis 1987). Pine (1993) popularised this concept and defined mass customisation as "providing tremendous variety and individual customization, at prices comparable to standard goods and services" to enable the generation of products and services "with enough variety and customization that nearly everyone finds exactly what they want". A more pragmatic definition was introduced by Tseng and Jiao (2001) who suggested that mass customisation corresponds to "the technologies and systems to deliver goods and services that meet individual customers' needs with near mass production efficiency" (Tseng and Jiao 2001).

Often, this definition is supplemented by the requirement that the individualised goods do not carry the price premiums associated traditionally with (craft) customisation (Davis 1987). However, consumers are frequently found to be willing to pay a price premium for customisation that reflects the increment of utility which customers gain from a product better fitted to their needs than the best standardised product attainable (Franke and Piller 2004; Franke et al. 2009). Hence, we opt for neither including a price proposition into the definition of mass customisation nor reducing the concept to 'on-demand manufacturing of lot sizes of one'. Custom products can be produced in larger quantities for an individual customer. This frequently happens in industrial markets, when, for example, a supplier provides a custom component that is integrated into a product. Indeed, one of the biggest lessons to impart is that there is no one best way to mass customise, and trying to copy successful companies such as Dell can lead to serious failures (Salvador et al. 2009). Take, for example, the widespread belief that mass customisation entails building products to order. This is not necessarily true. As discussed earlier, customers are looking for products that fit their needs, and they do not necessarily care whether those offerings are physically built to their order or whether those items come from a warehouse—just as long as their needs are fulfilled at a reasonable price (Brabazon and MacCarthy 2010). Consider again the example of Sears, a multibillion-dollar online business that uses avatars and style-matching technology to help customers browse through countless products, including kitchen appliances and furniture. Sears focuses on personalising the shopping experience, but not its products; and the results at some business units have been impressive: double-digit increases in the average order value. Or, consider again the example of Pandora Radio: every user of this internet radio station will praise its ability to customise an individual stream of music, different for any particular user. However, Pandora's custom music delivery system builds on matching 'standard' songs to a user's preferences, but not on customising the song itself.

However, to reap the benefits of mass customisation, managers must not think of it as a stand-alone business strategy for replacing production and distribution processes, but as a set of organisational capabilities that can enrich the portfolio of capabilities of their organisations. Mass customisation means to profit from the fact that all people are different, that is, turning heterogeneities in the customer domain into an opportunity to create value, rather than a problem to be minimised, challenging the 'one size fits all' assumption of traditional mass production.

4.4 Three Capabilities of Mass Customisation

Companies that are able to master a mass customisation regime have tended to build competences around a set of core capabilities. The key to mastering and profiting from mass customisation is to accept it as a set of organisational capabilities that can supplement and enrich an existing system. While specific answers on the nature and characteristics of these capabilities are clearly dependent on industry context or product characteristics, three fundamental groups of capabilities are suggested as being able to determine the ability of a business to mass customise. We call them 'solution space development', 'robust process design', and 'choice navigation' (the derivation of these capabilities builds on work by Salvador et al. 2008 and Salvador et al. 2009). These capabilities are briefly introduced in the following section and discussed in more detail in the remaining sections.

- *Solution Space Development:* First and foremost, a company seeking to embrace and adopt a policy mass customisation has to be able to understand the idiosyncratic needs of its customers. This is in contrast to the approach of a mass

producer, where the company focuses on identifying 'central tendencies' among its customers' needs, and targets them with a limited number of standard products. Conversely, a mass customiser has to identify the product attributes along which customer needs diverge the most. Once this is understood, the business knows what is required to properly cover the needs of its customers. Consequently, it can draw up the so-called solution space, clearly defining what it is going to offer and what it is not.
- *Robust Process Design:* A second critical requirement for mass customisation is related to the relative performance of the supply chain. Specifically, it is imperative that the increased variability in customers' requirements does not lead to significant deterioration in the company's operations' and supply chain performance (Pine et al. 1993). This demands a robust supply chain design—defined as the capability to reuse or re-combine existing organisational and supply chain resources to fulfill differentiated customers' needs. With robust process design, customised solutions can be delivered with near mass production efficiency and reliability.
- *Choice Navigation:* Finally, the business must be able to support customers in identifying their own problems and solutions, while minimising complexity and the burden of choice. When a customer is exposed to too many choices, the cognitive cost of evaluation can easily outweigh the increased utility from having more choices (Huffman and Kahn 1998; Piller 2005). As such, offering more product choices can easily prompt customers to postpone or suspend their buying decisions. Therefore, the third requirement is the organisational intelligence to simplify the navigation of the company's product assortment from the customers' perspective.

The methods behind these capabilities are often not new. Some of them have been around for many years. However, successful mass customisation demands the combination of these methods into capabilities in a meaningful and integrated way, to design a supply chain that creates value from serving individual customers differently. In the following, we will discuss the three fundamental capabilities of mass customisation in greater detail and also look into the approaches and practices connected with these capabilities.

4.4.1 Solution Space Development

A mass customiser must first identify the idiosyncratic needs of its customers, specifically, the product attributes along which customer needs diverge the most. This is in stark contrast to a mass producer, which must focus on serving universal needs, ideally shared by all the target customers. Once that information is known and understood, a business can define its 'solution space', clearly delineating what it will offer—and what it will not. This space determines the universe of benefits

4.4 Three Capabilities of Mass Customisation

an offer is intended to bring to customers and then within that universe which specific permutations of functionality can be provided (Pine 1995).

Options for customisation

From the perspective of product development, value through customisation can be achieved via three design features of a product (or service), any of which can become the starting point for customisation: the fit (measurements), the functionality, and the form (style and aesthetic design) of an offering (Piller 2005). These are generic dimensions that align the requirements of a customer with an offering. Along these dimensions, heterogeneities of demand can be derived from a customer perspective. The solution space should represent options for choice in those dimensions where customer heterogeneities are relevant.

Fit and comfort (measurements): The traditional starting point for addressing customisation in consumer goods' markets is to attempt to fit a product to the measurements provided by the client, for example, body measurements or the dimensions of a room or other physical object. Market research has identified better fit as one of the most persuasive arguments in favour of mass customisation. Often, however, it is a difficult attribute to satisfy, demanding complex systems to ascertain the customers' proportions exactly and to codify this information into a product which has to be based on a parametric design (for fulfilling the requirements of a stable solution space). This often calls for a total re-design of the product and the costly development of flexible product architectures with sufficient slack to accommodate all possible fitting requirements of the customer base. In sales, expensive 3D scanners or other devices are needed, which in turn demands highly qualified sales staff for their operation (Berger et al. 2005).

Functionality: Functionality addresses issues such as speed selection, precision, power, cushioning, output devices, interfaces, connectivity, upgradeability or similar technical attributes of an offering according to the requirements of the client. This is the traditional starting point for the process of customisation in industrial markets, where machines and equipment, for example, are adjusted, or components are produced according to the exact specifications of their buyers. Functionality demands similar efforts to elicit customer information about the desired individual function as the fit dimension. In manufacturing, however, the growing software content of many products is allowing the customisability of functional components to be more easily achieved.

Form (style and aesthetic design): This dimension concerns choices that relate to the senses, for example, selecting colours, styles, applications, cuts or flavours. Many mass customisation offerings in business-to-consumer (B2C) e-commerce are based on the possibility of co-designing the outer appearance of a product. This kind of customisation is often relatively straightforward to achieve in manufacturing, particularly if digital printing technology is appropriate as the means to provide the distinctive appearance of the product. The desire for a particular outer appearance is often inspired by fashions, peers and role models, and the individual's wish to mimic and adopt these trends. Along this line, the construct of 'consumers' need for uniqueness' has been discussed in the psychological marketing literature (Tepper et al. 2001). Consumers acquire and display material possessions for the purpose of feeling different to other people or they perform explicit actions in order to be recognised by others (counter-conformity motivation). Some consumers express their desire for uniqueness by selecting material objects (fashion) which are ahead of the average trend, by purchasing hand-crafted items, or vintage goods from non-traditional outlets. Mass customisation can be a further means to express their uniqueness, where consumers can design products according to their own personal specifications in order to look different from the rest.

To illustrate these options, consider the examples of shoes and cereals. With the former, fit is mostly defined by the last on which the shoe is formed, but also by the design of the uppers, insole and outsole etc. Style is an option for influencing the aesthetic design of the product, for example, the colour of the leather, or patterns. A shoe's functionality can be defined by its cushioning, heel form, or cleat structure. In the case of cereals, these options could be translated into package size (fit), taste (no chocolate and raisins, lots of strawberries), and nutrition (vitamins, special fibres).

Methods for solution space definition

To define the solution space, the company has to identify those needs where customers are different—and where they care about such differences. Matching the options represented by the solution space with the needs of the targeted market segment is a major success factor of mass customisation (Hvam et al. 2008). The core requirement at this stage is to access 'customer need information', that is, information about preferences, needs, desires, satisfaction, motives, etc. of the customers and users of the product or service offering. 'Need' information builds on an in-depth understanding and appreciation of the customers' requirements, operations and systems. Spotting untapped differences across customers is not a simple task, because information about customers' unfulfilled needs is "sticky"—that is, difficult to access and codify for the solutions provider (von Hippel 1998). While this problem is shared by both mass producers and mass customisers, it is more acutely felt by the latter due to the extreme fragmentation of customers' preferences. Understanding heterogeneous customer needs in terms of identifying differentiating attributes, validating product concepts, and collecting customer feedback can be a costly and complex endeavour, but several approaches can help.

The first is to engage in conventional market research techniques, that is, to meticulously gather data from representative customers on a chosen market sector. To reduce the risk of failure, need-related information from customers is integrated iteratively at many points in the new product development process (for example, Griffin and Hauser 1993; Dahan and Hauser 2002). The manufacturer selects and surveys a group of customers to obtain information on needs for new products, analyses the data, develops a responsive product idea, and screens this idea against customer preferences (needs) and purchasing decisions. This model is dominant particularly in the world of consumer goods, where market research methodologies such as focus groups, conjoint analysis, customer surveys, and analyses of customer complaints is used regularly to identify and evaluate customer needs and desires.

4.4 Three Capabilities of Mass Customisation

In particular, conjoint analysis, also called multi-attribute compositional analysis, can be regarded as a tool that is appropriate for supporting a company's solution space in a mass customisation environment. The term denotes a set of methods to measure and analyse consumers' preferences by assessing their perception of the value of various attributes of a product (Green et al. 1981; Green and Srinivasan 1990; Louviere 1994). The method is based on an experimental design that allows for systematically manipulating product or service descriptions shown to a respondent. In the conventional set-up, customers are asked to prioritise their choices by making trade-offs, which in turn reveal their view of utility values for each product alternative. A subset of possible combinations of product features is used to determine the relative importance of each feature in the purchasing decision. The respondents may be asked to arrange a list of combinations of product attributes in decreasing order of preference. Once this ranking is obtained, the utilities of different values of each attribute are identified based on the respondent's order of preference. This method is efficient in the sense that the survey does not need to be conducted using every possible combination of attributes. The utilities can be determined using a subset of possible attribute combinations. From these results, one can predict the desirability of the combinations that were not tested (Green et al. 1981). Some researchers have developed methods for solution space definition which build on a conjoint analysis methodology. Here, the approach is not used to identify the best product variants, but the options in a solution space that will be valued most by customers (Chen and Tseng 2005; Du et al. 2003; Siddique and Rosen 2003).

A second approach companies can use to define their solution space is to provide customers with toolkits for user co-creation (von Hippel and Katz 2002; Franke and Piller 2004). These are software design tools such as computer-aided design (CAD) systems, but with easy-to-use interfaces and libraries of basic modules and functionalities. With these toolkits, customers can, by themselves, translate their preferences directly into a product design, highlighting unsatisfied needs during the process. The resulting information can then be evaluated and potentially incorporated by the company into its solution space. When Fiat was developing its retro, award-winning Fiat 500, for example, the automaker created Concept Lab, an innovation toolkit that enabled customers to freely express their preferences regarding the interior of the car long before the first vehicle had been built. The company received more than 160,000 designs from customers—a product-development effort that no automaker could replicate internally. Subsequently, Fiat allowed people to comment on the submissions of others, providing an initial evaluation of these ideas. Mass producers can also benefit from innovation toolkits, but the technology is particularly useful for mass customisation, because it can be deployed at low-cost for large pools of heterogeneous customers.

Third, in developing their solution space, companies can employ some form of 'customer experience intelligence', that is, to apply methods for continuously collecting data on customer transactions, behaviours, or experiences and analysing that information to determine customer preferences. This also includes incorporating data not just from customers, but also from people who might have taken their business elsewhere. Consider, for example, information about products that someone has evaluated, but did not order. Such data can be obtained from log files generated by the browsing behaviour of people using online configurators (Rangaswamy and Pal 2003; Squire et al. 2004; Piller et al. 2004). By systematically analysing that information, managers can learn much about customer preferences, ultimately leading to a refined solution space. A company could, for instance, eliminate options that are rarely explored or selected, and it could add more choices for the popular components. In addition, customer feedback can even be used to improve the very algorithms that a particular application deploys. When someone skips a song that Pandora Radio has suggested, for example, that information is not just used to provide better personalisation of the music stream for that particular individual, it is also aggregated with similar feedback from millions of other customers to prevent the system from making that kind of incorrect recommendation in the future.

Modular product architectures

Once the relevant options to be represented in a solution space have been identified (an iterative, continuous process that will continue as long as the mass customisation offering exists), they have to be translated into a product architecture that will transfer these needs into solutions for the customer. It is important to note that mass customisation does not mean to offer limitless choice, but to offer choice that is restricted to options that are already represented in the fulfillment system. In the case of digital goods (or components), customisation possibilities may be infinite. In the case of physical goods, however, they are limited and are often represented by a modular product architecture.

Modularity is an essential element of any mass customisation strategy (Gilmore and Pine 1997; Duray 2002; Kumar 2005; Piller 2005; Salvador 2007). Each module serves one or more well-defined functions of the product and is available in several options that deliver a different performance level for the function(s) the product is intended to serve. This principle implies that mass customisation demands compromise: not all notional customisation options are being offered, but only those that are consistent with the capabilities of the processes, the given product architecture, and the given degree of variety.

The product family approach has been recognised as an effective means to accommodate an increasing product variety across diverse market niches while still being able to achieve economies of scale (Tseng and Jiao 2001; Zhang and Tseng 2007). In addition to leveraging the costs of delivering variety, a product family-based approach can reduce development risks by re-using proven elements in a company's activities and offerings. The backdrop of a product family is a well-planned architecture—the conceptual structure and overall logical organisation of generating a family of products—providing a generic umbrella to capture and utilise commonality. Within this architecture, each new product is instantiated and extends to anchor future designs to a common product line structure. The rationale of such a product family architecture lies not only in unburdening the knowledge base from keeping variant forms of the same solution, but also in modelling the design process of a class of products that can widely variegate designs based on individual customisation requirements within a coherent framework (Tseng and Jiao 2001). Setting the modular product family structure of a mass customisation system, and thus its solution space, becomes one of the foremost competitive capabilities of a mass customisation company.

4.4.2 Robust Process Design

A core idea of mass customisation is to ensure that an increased variability in customers' requirements will not significantly impair the company's operations and supply chain (Pine et al. 1993). This can be achieved through robust process design—the capability to re-use or re-combine existing organisational and supply chain resources to deliver customised solutions with high efficiency and reliability. Hence, a successful mass customisation system is characterised by stable, but still flexible, responsive processes that provide a dynamic flow of products (Pine 1995; Tu et al. 2001; Salvador et al. 2004; Badurdeen and Masel 2007). Value creation within robust processes is a major differentiating factor of mass customisation over conventional (craft) customisation. Traditional (craft) customisers re-invent not only their products, but also their processes for each individual customer. Mass customisers use stable processes to deliver high-variety goods (Pine et al. 1993), which allows them to achieve "near mass production efficiency", but it also implies that the customisation options are somehow limited. Customers are being served from a list of pre-defined options or components, the company's solution space.

Cost drivers of variety

The core objective of robust process design is to prevent or counterbalance the additional cost resulting from the flexibility a company needs to achieve in order to serve its customers individually. We can differentiate two sources of additional costs of flexibility (Su et al. 2005): (i) increased complexity and (ii) increased uncertainty in business operations, which by implication results in higher operational cost. A higher level of product customisation requires greater product variety, which in turn entails a greater number of parts, processes, suppliers, retailers, and distribution channels. A direct consequence of such proliferations is an increased complexity in managing all aspects of business from raw material procurement to production and eventually to distribution. Furthermore, an increase in product variety has the effect of introducing greater uncertainty in demand, increases in manufacturing cycle times, and increases in shipment lead times (Kumar and Piller 2006; Yao et al. 2007). Increased system complexity and uncertainties (in demand and lead time) drive the operational cost upward due to more complex planning, greater hedging, increased resource usage, more complex production setups, diseconomies of scope, and higher distribution cost spread throughout the supply chain. Finally, a sizeable increase in costs to offer choice navigation for customers is integral to a mass customisation strategy. This includes, for example, implementing a configuration system on a website or in a physical retail store.

Methods to establish robust processes

A number of different methods can be employed to reduce these additional costs, or even to prevent their occurrence at all. A primary mechanism to create robust processes in mass customisation is the application of delayed product differentiation (postponement) as discussed in Chap. 3. Postponement refers to partitioning the supply chain into two stages (Yang and Burns 2003; Yang et al. 2004). A 'standardised' portion of the product is produced during the first (upstream) stage of the supply chain, while the 'differentiated' portion of the product is produced in the second (downstream) stage of the chain, based on customer preferences which have been expressed via an order. The success of a postponement approach is a direct consequence of the fact that most companies offer a portfolio of products that consist of families of closely-related items which differ from each other in a limited number of differentiated features. An innovative example of postponement in the automotive industry would be to provide a standard version of the vehicle (a stripped or partially equipped version) to dealers and then allow the dealer to install, on the basis of customer-specific requests, options such as a CD/DVD player, the interior leather or fabric, and the cruise control system. Prior to the point of differentiation, product items are engineered so that as many parts or components of the products as possible are common to each configuration. Cost savings result from the risk-pooling effect and reduction in inventory holding costs (Yang et al. 2004). Additionally, as common performance levels of functionalities are selected by a number of customers, economies of scale can be achieved at the modular level for each version of the module, generating cost savings not available in pure customisation-oriented production systems.

While postponement starts at the design of the offerings, another possibility to achieve robust processes is through flexible automation (Tu et al. 2001; Zhang et al. 2003; Koste et al. 2004). Although 'flexible' and 'automation' might have been contradictory in the past, the two words together are now oxymoronic. In the automotive industry, for example, robots and automation are compatible with high levels of versatility and customisation. Even process industries (pharmaceuticals and chemicals' producers for example), generally regarded as environments where rigid automation and large batches are institutionalised, nowadays enjoy levels of flexibility once considered unattainable. Similarly, many intangible goods and services also lend themselves to flexible automated solutions often supported by the internet. In the case of the entertainment industry, increasing digitalisation is turning the entire product system over from the real to the virtual world.

A complementary approach to flexible automation is process modularity, which can be achieved by thinking of operational and supply chain processes as segments, each one linked to a specific source of variability in the customers' needs (Pine et al. 1993). As such, the company can serve different customer requirements by appropriately re-combining the process segments, without the need to create costly ad-hoc modules (Zhang et al. 2003). BMW's Mini factory, for instance, relies on individual mobile production cells with standardised robotic units. BMW can integrate the cells into an existing system in the plant within a few days, thus enabling the company to quickly adapt to unexpected swings in customer preferences without extensive modifications of its production areas. Process modularity can also be applied to service industries. IBM, for example, has been redesigning its consulting unit around configurable processes (called 'engagement models'). The objective is to fix the overall architecture of even complex projects while retaining enough adaptability to respond to the specific needs of a client.

To ensure the success of robust process designs, companies also need to invest in adaptive human capital (Bhattacharya et al. 2005). Specifically, employees and managers need to be creatively involved in process improvement (see Chap. 2) and be capable of dealing with novel and ambiguous tasks in order to offset any potential rigidity that is embedded in process structures and technologies. Moreover, machines and equipment are not capable of determining what a future solution space will look like. That task clearly requires managerial decision making, not software algorithms. Individuals require a broad knowledge base that stretches beyond their immediate functional specialism in order to be able to proficiently interact with other functions and support the process of identifying and delivering tailored solutions for the customer (Salvador et al. 2009). Such a broad knowledge base has to be complemented with open, relational attitudes that allow the individual to easily connect with other employees on an ad-hoc basis.

4.4.3 Choice Navigation

Lastly, a mass customiser must support customers in identifying their own needs and creating solutions while minimising complexity and the burden of choice. When a customer is exposed to a myriad of choice, the cost of evaluating the options can easily outweigh the additional benefit from having so many alternatives. The resulting syndrome has been called the "paradox of choice", (Schwartz 2004) in which too many options can actually reduce customer value instead of having the desired effect of increasing it (Huffman and Kahn 1998; Desmueles 2002). In such situations, customers might postpone their buying decisions or, worse, classify the vendor as difficult and undesirable. Recent research in marketing has addressed this issue in more detail and has found that the perceived cognitive cost is one of the highest hurdles towards a greater adoption of mass customisation from the consumer perspective (Dellaert and Stremersch 2005). To avoid this, companies have to provide the means of choice navigation to simplify the ways in which people explore their offerings.

Configuration Toolkits

The traditional approach for navigating the customer's choice in a mass customisation system has been to use product configuration systems, also referred to as "co-design toolkits" (Franke and Piller 2003, 2004). Co-design activities are performed in an act of company-to-customer interaction and co-operation (Khalid and Helander 2003; Tseng et al. 2003). As early as 1991, Udwadia and Kumar (1991) were envisioning customers and manufacturers becoming "co-constructors" (i.e., co-designers) of those products intended for each customer's individual use. In their view, co-construction would occur when customers had only a nebulous sense of what they wanted. Without the customers' deep involvement, the manufacturer would be unable to cater to each individualised product demand adequately. After this seminal publication, computer technology, particularly the capacity to simulate potential product designs before a purchase, has significantly aided a greater collaborative effort (Ulrich et al. 2003; Haug and Hvam 2007). This understanding represents one of the four forms of mass customisation as identified by Gilmore and Pine (1997), collaborative customisation. In this strategy, the manufacturer and customer work together to identify and satisfy a customer's needs via a system that allows easy articulation of exact requirements. Anderson-Connell et al. (2002) used the term "co-design" to describe a collaborative relationship between consumers and manufacturers in which, via a process of interaction between a design manager and a consumer, a product is designed according to consumer specification based on an existing portfolio of manufacturing capabilities.

In mass customisation, co-design activities are in general performed with the help of dedicated systems. These systems are known as configurators, choice boards, design systems, toolkits, or co-design platforms (Salvador and Forza, 2007; Hvam et al., 2008). They are responsible for guiding the user through the elicitation process. Whenever the term configurator or configuration system is quoted in the literature, for the most part, it is used in a technical sense, usually addressing a software tool. The success of such an interaction system, however, is by no means defined solely by its technological capabilities but also by its integration into the sales environment, its ability to foster learning, its ability to provide experience and process satisfaction, and its integration into the brand concept. Tools for user integration in a mass customisation system contain much more than arithmetic algorithms for combining modular components. Taking up an expression from von Hippel (2001), the more generic term "toolkits for customer co-design" might better describe the diverse activities taking place (Franke and Piller 2003). In a toolkit, different variants are represented, visualised, assessed, and priced with an accompanying learning-by-doing process for the user. The core idea is to engage customers into fast-cycle, trial-and-error learning processes (von Hippel 1998). As a result of this mechanism, customers can engage in multiple sequential experiments to test the match between the available options and their needs.

Choice navigation, however, does not just refer to preventing 'complexity of choice' and the negative effects of variety from the customers' perspective. Offering choice to customers in a meaningful way, on the contrary, can become a way for new profit opportunities (Franke and Schreier 2010). Recent research has shown that up to 50% of the additional willingness to pay for customised (consumer) products can be explained by the positive perception of the co-design process itself (Franke and Piller 2004; Schreier 2006; Franke and Schreier 2010; Merle et al. 2010). Product co-designs by customers may also provide symbolic (intrinsic and social) benefits, resulting from the actual process of co-design rather

4.4 Three Capabilities of Mass Customisation

than its outcome. Schreier (2006) quotes, for example, a pride-of-authorship effect. Customers may co-create something by themselves, which may add value due to the sheer enthusiasm about the result. This effect relates to the desire for uniqueness, as discussed before, but here it is based on a unique task and not the outcome. In addition to enjoyment, task accomplishment has a sense of creativity. Participating in a co-design process may be considered a highly creative problem-solving process by the individuals engaged in this task, thus becoming a motivator to purchase a mass customisation product.

An important prerequisite for customer satisfaction is derived from the ingenuity of the co-design process itself. The customer has to be capable of performing the task successfully. This competency issue concerns 'flow', a concept that has been used by researchers to explain how customer participation in a process increases satisfaction (Csikszentmihalyi 1990). Flow is the process of optimal experience achieved when motivated users perceive an appropriate balance between their skills and the task at hand during an interaction process (Novak et al. 2000). Interacting with a co-design toolkit may lead exactly to this state, as recent research in marketing has indicated (for example, Dellaert and Stremersch 2005; Franke et al. 2008; Fuchs et al. 2010). Accordingly, recent research has recommended several design parameters of a configurator that should facilitate this effect of process satisfaction (Randall et al. 2005; Dellaert and Dabholkar 2009; Franke et al. 2009).

The interaction between the manufacturer and the customer that is underlying a co-design process offers further possibilities for building loyalty and enduring customer relationships. Once a customer has successfully purchased a specific item, the knowledge acquired by the manufacturer represents a considerable barrier against any potential switching to other suppliers. Re-ordering becomes much easier for the customers. Consider the case of Adidas® (Berger et al. 2005). In 2001, the sportswear company introduced its mass customisation programme, 'mi adidas', offering custom sport shoes with respect to fit, functionality and aesthetics. The process starts with a customer who wishes to buy personalised running shoes at a price of approximately $150. The more customers tell the vendor about their likes and dislikes during the interaction process, the better the chance of a product being created that meets the customers' exact needs at the first attempt. After delivery of the customised product, feedback from the customer enhances Adidas's knowledge of that customer. The manufacturer can draw on detailed information about the customer for the next sale, ensuring that the service provided becomes quicker, easier, and more focused. The information status is increased and refined with each additional sale. This data is also used to suggest subsequent purchases automatically, once the life of the training shoes has ended (for Adidas® customers who exercise regularly and intensively, this can, in fact, be the case every few months).

When Adidas® enters such a relationship, it increases the revenue-generation potential from each customer, as, in addition to the actual product experience, it simplifies the purchasing decision, so that the customer keeps returning. Why would a customer switch loyalty to a competitor—even if one were available who could deliver a comparable customised product and experience—if Adidas® already has all the information necessary for supplying the product? A new supplier would need to start from scratch and repeat the initial process of collecting customer data. Moreover, the customer has now learned how the interaction and self-integration process can successfully result in the creation of a product. By aggregating information from a collection of individual customers, Adidas® also improves its market research knowledge. As a result, new products for the mass market can be planned more efficiently, and market research is more effective because of reliable access to data on market trends and customers' needs. This is of particular benefit to those companies that integrate large-scale make-to-stock production with tailored services. In this way, mass customisation can become an enabling strategy for mass production systems to become more efficient.

This learning relationship may lead to new cost-saving opportunities (Piller et al. 2004), based on better access to knowledge about the needs and demands of the customer base (Kotha 1995; Squire et al. 2004). This includes

- the reduced or eliminated need for forecasting product demand,
- reduced or eliminated finished goods' inventory,
- reduced product returns,
- reduced obsolescence or antiquated fashion risks, and
- the prevention of lost sales if customers cannot find the product in a retail store that fits their requirements.

The savings from these effects can be huge. Forrester Research estimated that the U.S. automotive industry could save up to $3,500 per vehicle by moving from its current build-to-stock model to a build-to-order system. Similarly, for the apparel industry, cost savings of up to 30% are estimated when switching to an on-demand system. Estimates for the apparel industry indicate that almost $300 billion are wasted annually due to erroneous forecasting, excessive inventory, fashion risks, and lost profits as a result of necessary discounts (Sanders 2005).

Advanced methods for choice navigation

The application of toolkits for customer co-design may be the most-commonly used approach to help customers to navigate choice in a mass customisation system. However, a number of other approaches also exist. One effective approach has been labeled "assortment matching" (Salvador et al. 2009), in which software automatically builds configurations for customers by matching models of their needs with characteristics of existing solution spaces (sets of options). Subsequently, customers only have to evaluate the pre-defined configurations, which saves considerable effort and time in the search process. Using special software, for example, customers at Sears.com can build avatars of themselves by selecting different body types, hair styles, facial characteristics and so on. From that information, the system can then recommend items out of the vast range of an online merchant. Alternatively, consider another example of assortment matching: Zafu.com has created a very profitable business model by taking body measurements of customers and then recommending the best-fitting pair of jeans from the existing assortments of many major brands. From their users' perspective, Zafu is offering a product that fits in a tailor-made fashion. However, from the fulfillment perspective, Zafu is simply matching standard inventory with individual needs. We consider this to be a true mass customisation strategy: providing all customers with what they want—at high efficiency.

Customers may not always be ready to make a decision after they have received recommendations. Perhaps they are uncertain about their real preferences, or perhaps the recommendations may not appear to fit their needs. In such cases, combining a recommendation system with a co-design toolkit is a pragmatic solution. Consider online shoppers at 121Time.com, a leading provider of mass-customised Swiss watches. Customers in the watch market may have a general idea of what they want, but while using an online configurator to play around with various options, combining colours and styles, they can actually see how one choice influences another and affects the entire appearance of a watch. Through such an iterative process, they learn about their own preferences—important information that is then represented in subsequent configurations.

A number of companies are engaging in even more innovative and ambitious approaches to choice navigation. The process has been completely automated in recent products that 'understand' how they should adapt to the user and then reconfigure themselves accordingly. Equipped with so-called 'embedded configuration capability', the products paradoxically are regarded as standard items for the manufacturer while the user experiences a customised solution (Piller et al. 2010). Such is the case with Adidas One®, a running shoe equipped with a magnetic sensor, used to adjust the cushioning and a microprocessor to control the process. When the shoe's heel strikes the ground, the sensor measures the amount of compression in its mid-sole and the microprocessor calculates whether the shoe is too soft or too firm for the wearer. A tiny motor then shortens or lengthens a cable attached to a plastic cushioning element, making it more rigid or pliable. With this system, the sport shoe is continuously adapting to different user needs during its use—without any customisation of the manufacturing process.

4.5 A Review of Mass Customisation

Mass customisation can be regarded as a response to today's heterogeneous demands and the need of companies to become truly customer centric. When judiciously designed and implemented, a mass customisation system provides across-the-board improvements in all aspects of operations and supply chain management: responsiveness, price, quality, and service (Ismail et al. 2007). Mass customisation is neither a 'one size fits all' approach nor the right strategy in all contexts. A recent survey by FedEx Corporation in the apparel industry found that more than 90% of the respondents agree that mass customisation will play a significant role in the next 5 years. Yet mass customisation initiatives have not always had successful outcomes. For example, Levi's Original Spin program (custom jeans) or Procter & Gamble's Reflect brand (custom cosmetics) are prominent examples of mass customisation efforts in blue-chip, multi-national enterprises that failed to fulfill their promise and were subsequently terminated.

We have suggested in this chapter that mass customisation requires a business to develop three fundamental capabilities. Admittedly, the development of these capabilities mandates for organisational changes that are often difficult, because of obdurate inertial forces that might exist within a company. We have seen a repeating pattern of companies that have failed in implementing mass customisation. These companies were unsuccessfully attempting to manage a process of change from a product-focused, mass producing mentality to one of customer-centricity (Moser 2007). Business managers and their employees become accustomed to an organisational logic shaped by the attitudes, behaviours, and assumptions that they have witnessed in their environments over a long time. The mindset of many managers remains conditioned by managerial routines, systems, and incentives created under a mass production way of doing things. However, shifting the locus of value creation toward true customer centricity requires no less than a radical change in the management mindset (Forza and Salvador 2007). Businesses must begin at the level of normative management with the challenge of changing the old and adversarial perceptions towards customers and developing an attitude of listening to and aligning with them. The introduction of a mass customisation initiative must always be preceded by a well-conceived and well-deliberated change management process that is directed towards making the organisation more customer-focused.

Furthermore, we believe that the obstacles to successful adoption can be overcome by using a variety of approaches, and that even small, incremental improvements can realise substantial benefits. It is important to remember that there is no one best way to mass customise: managers need to customise the approach to mass customisation thinking in ways that make the most sense for their specific businesses and supply chains. Businesses that have found productive means to implement methods and approaches to match the three capabilities typically will succeed in their mass customisation endeavours. Other businesses are 'ploughing a single furrow' to address just one of the capabilities. Mass customisation should be considered a journey (analogous to kaizen) rather than a destination. It is not about achieving a 'perfect' state of mass customisation (Salvador et al. 2009). What is important to most businesses is to continuously increase their overall capabilities to define the solution space, to design robust processes, and to help customers navigate through available choices. A business may already profit significantly from just implementing better, say, choice navigation capabilities to match diverse requests of customers not familiar with the product category. We have called this understanding "mass customisation thinking" (Piller and Tseng 2010). It provides a way to profit from recognising and addressing the heterogeneities of customers through institutionalising the three capabilities previously described and to apply them to designing a supply chain that creates value from serving customers individually.

4.6 Chapter Summary

Over the last decade, mass customisation has emerged as an effective approach to customer centricity, that is, to regard customers as individuals, to proactively develop products and services according to the individual customer's preferences, and to efficiently produce and distribute these offerings. Put simply, the goal of mass customisation is to efficiently provide customers what they want, when they want it. This chapter discussed the background of mass customisation and its underlying fundamental capabilities: solution space definition, the design of robust processes, and choice navigation. The conclusion is that a business should customise its mass customisation strategy based on the requirements of its customer base, the state of its competition, and the technology available. It should not blindly use successful mass customisers as templates to copy. Mass customisation can be a powerful source of sustainable competitive advantage, it is an entrepreneurial endeavour that is broadly applicable to any business for which customers might feasibly be willing to pay for tailored solutions or experiences. It can be seen as a strategic mechanism to provide solutions to align the organisation and the supply chain with customers' needs.

References

Anderson C (2007) The long tail: how endless choice is creating unlimited demand. Random House, London

Anderson-Connell LJ, Ulrich PV, Brannon EL (2002) A consumer-driven model for mass customization in the apparel market. J Fash Mark Manag 6(3):240–258

Badurdeen F, Masel D (2007) A modular minicell configuration for mass customization manufacturing. Int J Mass Cust 2(1/2):39–56

Berger C, Moeslein K, Piller F, Reichwald R (2005) Co-designing the customer interface for customer-centric strategies: Learning from exploratory research. Eur Manag Rev 2(3):70–87

Bhattacharya M, Gibson DE, Doty H (2005) The effects of flexibility in employee skills, employee behaviors, and human resource practices on firm performance. J Manag 31(4):622–640

Brabazon PG, MacCarthy B (2010) On Markovian approximations for virtual-build-to-order systems. J Oper Res Soc 61(10):1471–1484

Chamberlin EH (1962) The theory of monopolistic competition: a re-orientation of value theory. Harvard University Press, Cambridge

Chen S, Tseng M (2005) Defining specifications for custom products: a multi-attribute negotiation approach. CIRP Ann 54(1):299–304

Csikszentmihalyi M (1990) Flow: the psychology of optimal experience. Harper & Row, New York

Dahan E, Hauser J (2002) The virtual customer. J Prod Innov Manag 19(5):332–353

Davis S (1987) Future perfect. Addison-Wesley, Reading

Day GS (1994) The capabilities of market-driven organization. J Mark 58(10):37–52

Dellaert BG, Dabholkar P (2009) Increasing the attractiveness of mass customization: the role of complementary online services and range of options. Int J Electron Commer 13(2):43–70

Dellaert BGC, Stremersch S (2005) Marketing mass customized products: Striking the balance between utility and complexity. J Mark Res 43(2):219–227

Desmueles R (2002) The impact of variety on consumer happiness: Marketing and the tyranny of freedom. Acad Mark Sci Rev 12:1–33

Drucker PF (1954) The practice of management. Harper & Row, New York

Du X, Jiao J, Tseng M (2003) Identifying customer need patterns for customization and personalization. Integr Manuf Sys 14(5):387–396

Duray R (2002) Mass customization origins: mass or custom manufacturing? Int J Oper Prod Manag 22(3):314–330

Forza C, Salvador F (2007) Product information management for mass customization: connecting customer front-end and back-end for fast and efficient personalization. Palgrave Macmillian, London

Franke N, Piller F (2003) Key research issues in user interaction with configuration toolkits in a mass customization system. Int J Technol Manag 26:578–599

Franke N, Piller F (2004) Toolkits for user innovation and design: an exploration of user interaction and value creation. J Prod Innov Manag 21(6):401–415

Franke N, Keinz P, Schreier M (2008) Complementing mass customization toolkits with user communities: how peer input improves customer self-design. J Prod Innov Manag 25(6):546–559

Franke N, Keinz P, Steger C (2009) Testing the value of customization: when do customers really prefer products tailored to their preferences? J Mark 73(5):103–121

Franke N, Schreier M (2010) Why customers value self-designed products: the importance of process effort and enjoyment. J Prod Innov Manag 27(7):1020–1031

Fuchs C, Schreier M, Prandelli E (2010) The psychological effects of empowerment strategies on consumers' product demand. J Mark 74(1):65–79

Gilmore JH, Pine BJ (1997) The four faces of mass customization. Harv Bus Rev 75(1):91–101

Green P, Carroll J, Goldberg S (1981) A general approach to product design optimization via conjoint analysis. J Mark 43:17–35

Green P, Srinivasan V (1990) Conjoint analysis in marketing research. J Mark 54(4):3–19

Griffin A, Hauser J (1993) The voice of the customer. Mark Sci 12(1):1–27

Haug A, Hvam L (2007) Modeling techniques of a documentation system that supports the development and maintenance of product configuration systems. Int J Mass Cust 2(1/2):1–18

Homburg C, Workman JP, Jensen O (2000) Fundamental changes in marketing organization: the movement toward a customer-focused organizational structure. Acad Mark Sci J 28(4):459–478

Huffman C, Kahn B (1998) Variety for sale: Mass customization or mass confusion? J Retail 74(4):491–513

Hvam L, Mortensen N, Riis J (2008) Product customization. Springer, New York

Ismail H, Reid IR, Mooney J, Poolton J, Arokiam I (2007) How small and medium enterprises effectively participate in the mass customization game. IEEE Trans Eng Manag 54(1):86–97

Kaplan AK, Schoder D, Haenlein M (2007) Factors influencing the adoption of mass customization: The impact of base category consumption frequency and need satisfaction. J Prod Innov Manag 24(2):101–116

Khalid HM, Helander MG (2003) Web-based do-it-yourself product design. In: Tseng M, Piller F (eds) The customer centric enterprise: advances in mass customization and personalization. Springer, New York, pp 247–266

Koste LL, Malhotra MK, Sharma S (2004) Measuring dimensions of manufacturing flexibility. J Oper Manag 22(2):171–196

Kotha S (1995) Mass customization: implementing the emerging paradigm for competitive advantage. Strateg Manag J 16:21–42

Kotler P (1991) Marketing management. Prentice-Hall, Englewood-Cliffs (Original work published 1967)

Kumar A (2005) Mass customization: metrics and modularity. Int J Flex Manuf Sys 16(4):287–312

References

Kumar A, Piller F (2006) Mass customization and financial services. J Financial Transform 18(3):125–131
Lampel J, Mintzberg H (1996) Customizing customization. Sloan Manag Rev 38(1):21–30
Louviere J (1994) Conjoint analysis. In: Bagozzi R (ed) Advanced methods of marketing research. Blackwell, Cambridge, pp 223–259
McKenna R (1991) Relationship marketing. Addison-Wesley, Reading
Merle A, Chandon J, Roux E, Alizon F (2010) Perceived value of the mass-customized product and mass customization experience for individual consumers. Prod Oper Manag 19(5):503–514
Moser K (2007) Mass customization strategies: development of a competence-based framework for identifying different mass customization strategies. Lulu, New York
Narver JC, Slater SF (1990) The effect of a marketing orientation on business profitability. J Mark 54(10):20–35
Novak T, Hoffmann D, Yung Y (2000) Measuring the customer experience in online environments: a structural modeling approach. Mark Sci 19(1):22–42
Parasuraman A, Grewal D (2000) Serving customers and consumers effectively in the twenty-first century. Acad Mark Sci J 28(1):9–16
Piller F (2005) Mass customization: reflections on the state of the concept. Int J Flex Manuf Sys 16(4):313–334
Piller F (2008) Mass customization. In: Wankel C (ed) The handbook of 21st century management. Sage Publications, Thousand Oaks, pp 420–430
Piller F, Ihl C, Steiner F (2010) Embedded toolkits for user co-design: A technology acceptance study of product adaptability in the usage stage. Proceedings of the 43th Hawaii International Conference on System Science (HICSS) 43(1):1–10
Piller F, Möslein K, Stotko C (2004) Does mass customization pay? An economic approach to evaluate customer integration. Production planning & control 15(4):435–444
Piller F, Reichwald R, Tseng M (2006) Competitive advantage through customer centric enterprises. Int J Mass Cust 1(2/3):157–165
Piller F, Tseng M (2010) Mass customization thinking: moving from pilot stage to an established business strategy. In: Piller F, Tseng M (eds) Handbook of research in mass customization and personalization, part 1: strategies and concepts. World Scientific Publishing, New York and Singapore, pp 1–18
Pine BJ (1993) Mass customization. Harvard Business School Press, Boston
Pine BJ (1995) Challenges to total quality management in manufacturing. In: Cortada JW, Woods JA (eds) The quality yearbook. McGraw-Hill, New York, pp 69–75
Pine BJ, Victor B, Boynton AC (1993) Making mass customization work. Harv Bus Rev 71(5):108–119
Randall T, Terwiesch C, Ulrich K (2005) Principles for user design of customized products. Calif Manag Rev 47(4):68–85
Rangaswamy A, Pal N (2003) Gaining business value from personalization technologies. In: Pal N, Rangaswamy A (eds) The power of one: gaining business value from personalization technologies. Trafford Publishing, Victoria, pp 1–9
Salvador F (2007) Towards a product modularity construct: literature review and reconceptualization. IEEE Trans Eng Manag 54(2):219–240
Salvador F, de Holan M, Piller F (2009) Cracking the code of mass customization. MIT Sloan Manag Rev 50(3):70–79
Salvador F, Forza C (2007) Principles for efficient and effective sales configuration design. Int J Mass Cust 2(1/2):114–127
Salvador F, Rungtusanatham M, Forza C (2004) Supply-chain configurations for mass customization. Prod Plan Control 15(4):381–397
Salvador F, Rungtusanatham M, Akpinar S, Forza C (2008) Strategic capabilities for mass customization: theoretical synthesis and empirical evidence. Academy of Management Best Paper Proceedings

Sanders F (2005) Financial rewards of mass customization. Proceedings of the 2005 world congress on mass customization and personalization. Hong Kong University of Science and Technology, Hong Kong

Schreier M (2006) The value increment of mass-customized products: an empirical assessment. J Consumer Behav 5(4):317–327

Schwartz B (2004) The paradox of choice: why more is less. Ecco, New York

Sheth JN, Sisodia RS, Sharma A (2000) The antecedents and consequences of customer-centric marketing. J Acad Mark Sci 28(1):55–66

Siddique Z, Rosen D (2003) Common platform architecture: identification for a set of similar products. In: Tseng M, Piller F (eds) The customer centric enterprise: advances in mass customization and personalization. Springer, New York, pp 163–182

Smith WR (1956) Product differentiation and market segmentation as alternative marketing strategies. J Mark 21(3):3–8

Squire B, Readman J, Brown S, Bessant J (2004) Mass customization: the key to customer value? Prod Plan Control 15(4):459–471

Su JCP, Chang Y, Ferguson M (2005) Evaluation of postponement structures to accommodate mass customization. J Oper Manag 23(3/4):305–318

Tepper K, Bearden WO, Hunter GL (2001) Consumers' need for uniqueness: Scale development and validation. J Consumer Res 28(1):50–66

Tseng M, Jiao J (2001) Mass customization. In: Salvendy G (ed) Handbook of industrial engineering. Wiley, New York, pp 684–709

Tseng MM, Jiao RJ, Wang C (2010) Design for mass personalization. CIRP Ann Manuf Technol 59(1):175–179

Tseng M, Kjellberg T, Lu S (2003) Design in the new e-commerce era. Ann CIRP 52(2):509–519

Tseng M, Piller F (2003) The customer centric enterprise. In: Tseng M, Piller F (eds) The customer centric enterprise: advances in mass customization and personalization. Springer, New York, pp 1–18

Tu Q, Vonderembse MA, Ragu-Nathan TS (2001) The impact of time-based manufacturing practices on mass customization and value to customer. J Oper Manag 19(2):201–217

Udwadia FE, Kumar R (1991) Impact of customer co-construction in product/service markets. Int J Technol Forecast Soc Chang 40:261–272

Ulrich P, Anderson-Connell L, Wu W (2003) Consumer co-design of apparel for mass customization. J Fash Mark Manag 7(4):398–412

von Hippel E (1998) Economics of product development by users: the impact of "sticky" local information. Manag Sci 44(5):629–644

von Hippel E (2001) Perspective: user toolkits for innovation. J Prod Innov Manag 18(4):247–257

von Hippel E, Katz R (2002) Shifting innovation to users via toolkits. Manag Sci 48(7):821–833

Yang B, Burns ND (2003) Implications of postponement for the supply chain. Int J Prod Res 41(9):2075–2090

Yang B, Burns ND, Backhouse CJ (2004) Postponement: a review and an integrated framework. Int J Oper Prod Manag 24(5):468–487

Yao S, Han X, Yang Y, Rong Y (2007) Computer-aided manufacturing planning for mass customization. Int J Adv Manuf Technol 32(1/2):194–204

Zhang M, Tseng M (2007) A product and process modeling based approach to study cost implications of product variety in mass customization. IEEE Trans Eng Manag 54(1):130–144

Zhang Q, Vonderembse MA, Lim J-S (2003) Manufacturing flexibility: defining and analyzing relationships among competence, capability, and customer satisfaction. J Oper Manag 21(2):173–191

Zuboff S, Maxmin J (2003) The support economy. Viking Penguin, London

Chapter 5
Network Collaboration: Vertical and Horizontal Partnerships

5.1 Introduction

Supply chain structures differ significantly according to the nature of the product supplied and the type of industrial sector served. Some supply chains, because of the complexity of the product or a strategic desire to outsource non-core activities and source globally, are intricate with multiple tiers and branches; others, by virtue of the relative simplicity of the product, or where strategy has led to an organisational logic that requires the retention of a large proportion of manufacturing or business processes, possess a high degree of vertical integration, and, therefore, are often shallower, narrower and less complex. Some supply chains are characterised by convoluted logistics structures, some are configured to sell products to order rather than from stock, some are designed to produce, transport and sell the same product through multiple channels while others are dedicated to a single channel. In all, students of supply chain management are required to master a quite diverse array of network design options.

For example, aerospace supply chains are relatively slow moving and structurally very deep whereas the fast moving consumer goods (FMCG) industry, concerned with the provision of retail products such as food items, toiletries, and household cleaning products operate in highly competitive environments, make-to-replenish, and have supply chain configurations that are relatively narrow and inventory-centric. Organisations operating in process industries have a set of different characteristics. Chemicals production is an archetypal process industry. Julka et al. (2002) noted that chemical industry supply chains are often long and their plant configurations are rigid. Furthermore, the supply chain can represent as much as 60–80% of a typical chemical manufacturer's costs. As a partial consequence of these traits, despite the fact that some process industry supply chains are improving their flexibility, supply innovations and solutions, habitual in other industries, have had limited penetration in many chemical industry supply arrangements.

The automotive sector is a popular choice of study environment for academics and industry analysts alike. The business and academic literature is replete with automotive commentaries, analyses and cases. Economists have often regarded the sector as a barometer of economic health, whilst others have studied the sector for the extensive complexity and variability of its highly choreographed operations and supply networks. A typical automotive supply network is a fast-moving, 'high-energy' environment with an end product that is subject to frequent updates, is produced in high volume, consists of thousands of parts and modules, and is highly customisable. The complexity of automotive products, the intensity of competition and the volume of demand have led to the automotive industry having some of the most sophisticated supply chain arrangements found in any sector.

5.2 Outsourcing and Supplier Parks

5.2.1 Introduction

Organisations are acutely aware of the fundamental business imperative to retain in-house the expertise associated with those activities that are key to maintaining their competitive advantage (refer to Sect. 1.3, 'The Make-orBuy Decision'). Outsourcing non-core activities results in a greater reliance on a supplier's expertise and concentrating on core competence is seen as a way to transform fixed into variable costs. In an intensively competitive milieu such as the automotive sector where being lean is a key priority, outsourcing provides businesses with the opportunity to reduce supply chain costs. For many organisations in manufacturing, outsourcing has become a routine decision within the process of business strategy formulation. The outsourcing process results in a shift of the external boundary of the organisation through the re-allocation of responsibility for carrying out tasks from inside the organisation to an external organisational entity. The process of outsourcing is often gradual and undertaken incrementally over a number of projects. Organisations sometimes progress from outsourcing relatively simple processes to more sophisticated modules and activities. Automakers, for example, may at first outsource a relatively simple operation such as wheel and tyre assembly before graduating to more complex processes and more high-value modules. Wheel and tyre assembly can be seen as a module that is relatively low in complexity and value. An instrument panel/cockpit assembly is an example of a module that is high in complexity and high in value.

The supplier park business model is a consequence of outsourcing. Typical activities carried out by supplier parks include warehousing and inventory management, sequencing, assembly and late configuration. According to Larsson (2002), a supplier park maximises the reliability of supply as the transport time from finished component to assembly is no more than a few minutes. Larsson (2002) also highlighted that it is modular production that is often cited as one of

5.2 Outsourcing and Supplier Parks   97

the major motivations behind the organisational and location logic of supplier parks. Frigant and Lung (2002) noted that a supplier's proximity to production preserves the coherency between the external transfers of goods and the internal rhythms of production. Also the researchers highlight that proximity permits a close articulation between a product's physical flows and the information flows that are associated with the delivery process.

Supplier parks can also be regarded as a consequence of the close relationships that exist between OEMs and their suppliers. They are also major sources of economic impact to the regions in which they are located. The automotive supplier park model typically comprises a vehicle assembly plant and a number of facilities operated by suppliers or third-party logistics providers (3PLs) which are situated in close proximity to the assembly plant. The suppliers' production facilities on an automotive supplier park are totally dedicated to the manufacturing and assembly of components or subsystems which are delivered just-in-time, or in sequence to the vehicle final assembly line. This can be looked upon as a synchronous supply arrangement. The configuration of supply, production and logistics capabilities found in automotive supplier parks creates a tight, integrated cluster that results in highly effective co-ordination and major advantages in terms of inventory and transportation costs. The development of many supplier parks in Western Europe has been possible through OEM investments and funding from the European Union and local government grants.

In the supplier park model, the close proximity between suppliers and the OEM may be seen as a disadvantage by some. One characteristic of most suppliers on supplier parks is that they are dedicated to only one vehicle plant: the OEM assembly operations consume all of the supplier's output. In this scenario, the supplier's production facility becomes totally reliant on providing components or subsystems just-in-time or just-in-sequence to the OEM's assembly lines.

5.2.2 *The Implications of Modularity*

The widespread adoption of modularity in the automotive industry has supported the rationale, planning and execution of the supplier park model. Vehicle modules can be complex and valuable subsystems (for example, seats and instrument panels), thus the need for suitable production and logistical solutions becomes particularly important. Moreover, in the automotive industry it is possible to appreciate that modular production and sequenced delivery can be seen as mutually reinforcing in the creation of supplier parks. Locating a business unit close to the customer creates a short supply line and allows for efficient synchronicity of production and delivery while at the same time taking advantage of the entire organisation network to gain economies of scale by centralising design and purchasing (Larsson 2002).

Frigant and Lung (2002) recognised two different types of automotive industrial cluster. These are termed modular consortia and industrial condominia.

In a modular consortium, the OEM has no direct involvement in manufacturing. Instead, first-tier suppliers act as 'supplier implants' and are responsible for the preparation and assembly of modules on the vehicle assembly line. An example of a modular consortium is the Smart production facility in Hambach, France. Suppliers to the site produce modules that include the powertrain, doors, cockpits and plastic body panels.

In an industrial condominium, the production of subsystems is assigned to suppliers located in the immediate vicinity of the assembly plant while the OEM retains control over the final vehicle assembly process. This is the most common type found in Western Europe. Examples of industrial condominia include the supplier parks of Martorell, Spain (Seat), Valencia, Spain (Ford), and Halewood, UK (Jaguar Land Rover). Supplier parks in the form of industrial condominia come in different sizes. For example, Martorell in Spain has an annual capacity of more than 550,000 units whilst Halewood in England has a capacity of approximately 100,000 units per year.

5.2.3 Physical Configurations of Supplier Parks

In the industrial condominium supplier park, the proximity of suppliers to the OEM provides supply chain planners with the opportunity of having different types of design configuration. In one type of configuration, contiguity can be achieved through the existence of overhead tunnels comprising conveyor systems. Figure 5.1 depicts a cluster of suppliers located adjacent to the OEM site. Inside each overhead tunnel there is a conveyor system used to transport the finished components and modules to the point-of-fit at the OEM site. The implication is that first-tier suppliers carry no finished goods' inventory. Also, the OEM does not carry inventory of the components and modules received. An example of this configuration is the supplier park located in Valencia, Spain. The Valencia site has a capacity of over 350,000 units a year. Conveyor systems are used to transport modules and components from suppliers direct to the points-of-fit at the vehicle assembly plant.

In the Valencia supplier park, first-tier suppliers and a service provider responsible for sequencing and assembly operations operate conveyor systems that transfer components and modules directly to the vehicle assembly line from the suppliers' facilities. Those supplier facilities without conveyors connecting them to the OEM assembly line are typically served by small trailers doing rounds around the perimeter of the supplier park. Figure 5.2 depicts the configuration of a conventional automotive supplier park arrangement.

In the scenario portrayed by Fig. 5.2, components and modules produced by the first-tier suppliers are typically delivered within a short time frame from the beginning of the first-tier's assembly operation. Examples of items produced and delivered in sequence using this configuration from the supplier to the vehicle assembly line include seat-sets, instrument panels and bumpers. The total time from the broadcast of the final assembly sequence to the suppliers and their

5.2 Outsourcing and Supplier Parks

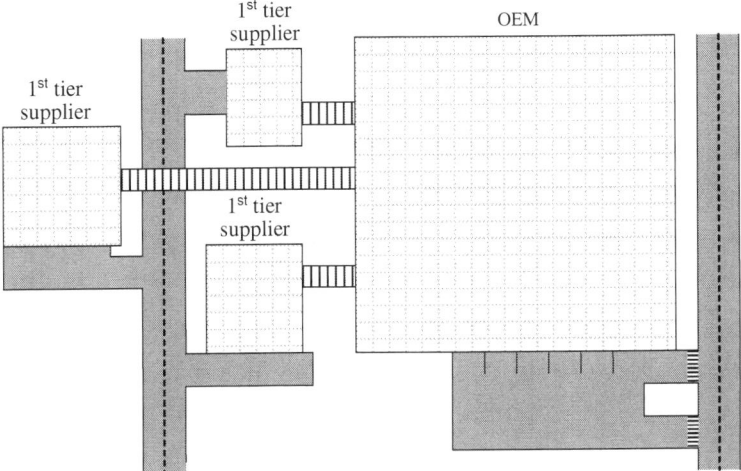

Fig. 5.1 Supplier park with overhead conveyors

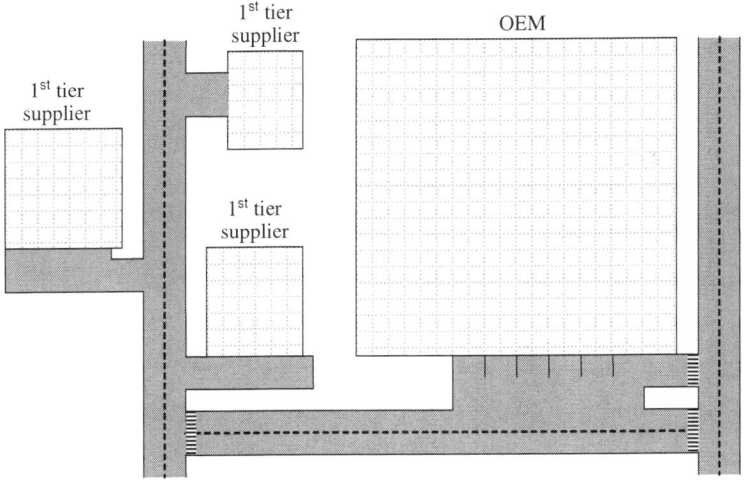

Fig. 5.2 Supplier park with linking roads

assembly and delivery of items to the vehicle assembler determines the response time available to suppliers. Items are typically placed in a cart in sequence, and delivered to the point-of-fit at the OEM.

The supplier park is a complex agglomeration of production and logistics activities that provides automotive managers with some challenges. Although the supplier park model has proven to be a very successful model at significantly reducing inventory levels and facilitating the adoption of synchronised deliveries of components and modules, a downside is that there are often considerable start-up investments required. For example, DaimlerChrysler invested around $2.1 billion USD at a supplier park in Toledo, Ohio, USA and its neighbouring Toledo North Assembly Plant. The three key suppliers to the park, Magna Steyr (operator of the paint shop), Hyundai Mobis (chassis assembler) and Kuka (operator of the body shop) also made significant investments.

On the other hand, several factors may work against the implementation of supplier parks. One of them is economic uncertainty. For example, in 2001 DaimlerChrysler had to close a plant with a supplier park configuration just after 19 months in operation. The Campo Largo, Brazil plant which built pickup trucks had to close down as a result of the economic uncertainty affecting the country at that time. It is not only the OEM that is disadvantaged if things happen to go awry with the commissioning and management of a supplier park. For a supplier park to succeed, the OEM has to take into account that although suppliers are usually given exclusivity to provide specific modules and components, the suppliers' operations must be profitable. This is important in order to secure the long-term viability of the supplier park. Furthermore, if there are problems, this can be compounded by further interruptions if there is a constant substitution of suppliers.

5.2.4 Proximate Supply and Vertical Integration

Organisations can be frustrated at the lack of responsiveness of their supply chains to customers' needs. One dynamic business environment is the textile and garments industry. This sector is highly competitive and in recent years only a handful of organisations have been capable of growing both sales and market share. Outsourcing production to factories in the far-east, principally mainland China, is a trend that has become the norm in the industry. It is not unusual to find organisations who own no production facilities at all.

However, there are examples within the sector of organisations that have designed highly-responsive supply chains and have used the agile supply business model as a source of competitive advantage. One such example is Zara. Zara, Inditex's flagship brand, is one of Europe's most well-known names in the fashion apparel sector. Zara's retail stores can be found in Japan, China, in numerous capital and other cities in Europe, and in the Americas, from Canada to Argentina. Zara customers are usually fashion-conscious, young people living in cities.

One conspicuous characteristic of Zara is its level of vertical integration. The high degree of vertical integration in the supply chain is a key contributing factor for Zara being able to operate a responsive supply chain. According to Ferdows et al. (2004), Zara can design, produce and put on display at its retail stores new clothing ranges in just 15 days. Other companies within the sector typically fall considerably short of such a level of performance. Much of the short lead time in the supply chain is attributable to the fact that Zara owns and operates a number of factories next to its headquarters in La Coruña, north east Spain, two in Catalonia and one in Lithuania. The company uses its own facilities to manufacture its complex designs often outsourcing the simpler ones to external producers. Zara produces in small batches and once an order has been fulfilled, it is sent to Zara's distribution centre and from there is dispatched to retailers, usually overnight. This implies that garments are shipped by truck to European retail stores or by plane for stores in the Americas and Asia. The lead time of 15 days includes the stages of design, fabric production, garment production, distribution and retail display. One key characteristic of Zara's operations is the availability of additional 'surge' capacity at its factories.

It is important to acknowledge the quality of the information flow present in the Zara supply chain. Every stage in the supply chain receives and transmits information to the other stages, thus data is constantly updated. This glass pipeline effect of transparent access of information ensures that the bullwhip effect is mitigated, so small disturbances do not get amplified and, as a consequence, disrupt the entire supply chain.

The business model pioneered by Zara is unusual because it has become a popular trend in the textile industry to outsource production. Hence, retailers are typically required to keep relatively high levels of inventory to protect themselves against stock-outs due to the long transit times associated with garment manufacturing being undertaken a long distance away. The lead time associated with supply chains that follow a business model based on outsourcing business operations is longer compared to Zara's. Outsourcing garment production to the far-east involves moving containers by sea. Usually, the time it takes a container to reach Europe from China can be up to four weeks. Once it reaches the distribution centre in Europe, it can take a further 48 h to reach the retail store in the country of destination.

Ferdows et al. (2004) examined Zara's supply chain and highlighted a series of characteristics. These are summarised as follows:

- *Highly-responsive supply chain.* The company can design, produce, and deliver a new garment and put it on display in its retail stores worldwide within a short space of time. Zara offers a large variety of the latest designs in very short periods of time and in limited quantities. Also the company is able to collect a significant proportion of the full ticket price on its retail clothing sales.
- *In-house production.* Zara has retained some production and intentionally leaves extra capacity. Zara has made major capital investments in production and distribution facilities which have contributed to its responsiveness to new and fluctuating demands. Zara's factories are garment-specialised and by using sophisticated JIT systems, the company is able to customise its processes and exploit innovations. Zara uses 'postponement' to gain more speed and flexibility, purchasing more than 50% of its fabrics undyed so that it can react faster to mid-season colour changes.
- *Supply chain ownership.* Zara is vertically integrated. Instead of relying on outside partners, the company manages the design, warehousing, distribution, and logistics functions itself.

- *Information flow.* Zara's supply chain is glass pipeline-enabled and organised to communicate information quickly and reliably to Zara's designers and production staff.
- *Tracking materials and products.* Zara is able to track materials and products in real time at every stage including products on display in the retail stores. It has been able to create a fast, closed information loop between the end users and the upstream operations of design, procurement, production, and distribution.

5.3 Horizontal Partnerships

Vertical collaboration is concerned with partnerships formed along a linear, upstream-downstream supply chain continuum. Conventional customer-supplier relationships are vertical in nature and this text, so far, has primarily concentrated on vertical alliances. Horizontal collaboration, however, is a growing part of supply chain design. It concerns the collaboration between organisational entities providing the same or similar service. It has been formally defined as "a business agreement between two or more companies at the same level in the supply chain or network in order to allow ease of work and co-operation towards achieving a common objective" (Bahinipati et al. 2009). The implication is that horizontal collaboration can concern the formation of productive partnerships between competing as well as non-competing organisations (Naesens et al. 2009).

The execution of horizontal collaboration can take many forms. These include aggregated procurement, shared services and joint product design. However, such collaborations are not organic and do not ordinarily happen without specific, innovation-driven ambitions. Lydeka and Adomavicius (2007) identified four main forms of horizontal collaboration in the logistics industry. The first concerned collaboration between partners that leads to direct cost reductions. Leveraging purchasing power through collaborative procurement is an appropriate example of this first form of horizontal alliance. The second was termed "consortium" and referred to projects where companies co-operate to serve customers larger than they could each serve individually. The third form concerned the sharing of information in an attempt to solve common problems and the fourth concerned the creation of new business ventures between the collaborators.

5.3.1 Horizontal Collaboration in Logistics

A survey (Everington et al. 2010) was recently undertaken in conjunction with the authors in order to explore and gain an understanding of collaborative practices and approaches to horizontal collaboration being undertaken by companies operating within the logistics sector. Figure 5.3 illustrates the degree of active participation and interest in horizontal collaboration from the 205 survey respondents.

5.3 Horizontal Partnerships

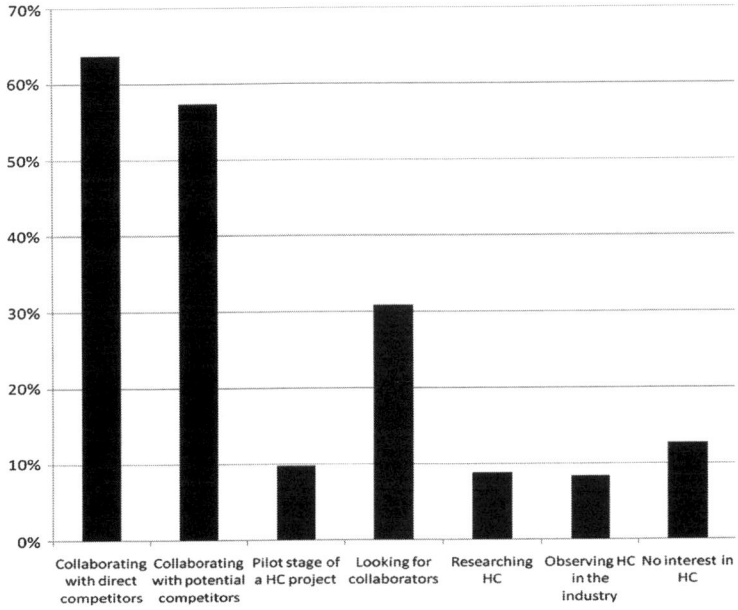

Fig. 5.3 Logistics providers' involvement in horizontal collaboration

Respondents were from a range of logistics providers with a range of different organisation sizes. 83% of the respondents indicated that their company participates in some form of horizontal collaboration.

Figure 5.4 depicts the drivers of horizontal collaboration identified by the respondents. Reducing transport costs was found to be the primary driver for horizontal collaboration with accessing new markets, enhancing customer service and improving vehicle fill utilisation being other sources of motivation. Figure 5.5 depicts the perceived barriers to horizontal collaboration. Lack of trust and fear of competitors accessing sensitive information were identified as the principal barriers. Relating company size to the perceived barriers to collaboration, several trends were identified. In particular, 'lack of management support' and 'limited precedence of similar initiatives' were found to be conspicuous barriers to collaboration for the smaller logistics providers that took part in the survey. However, 70% of companies with an annual turnover of less than £5 million are involved in some form of horizontal collaboration suggesting that, although there are barriers to collaboration, they can usually be overcome.

The most common mode of collaboration (refer to Fig. 5.6) that was found from the study concerned the sharing of services, with the consolidation of freight (complementary) also significant. Over half of the companies surveyed which are collaborating horizontally are involved in the consolidation of freight and the sharing of services. It was also noted that sharing services is particularly conspicuous with the smaller providers whereas consolidation of freight flows (non-

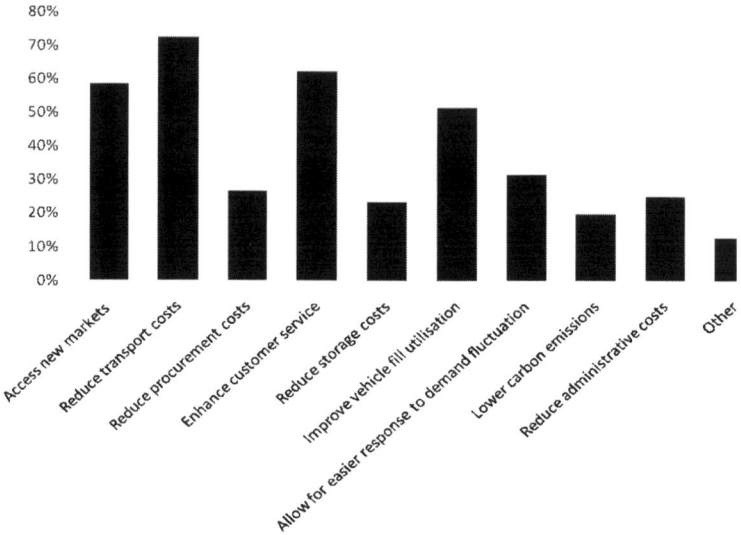

Fig. 5.4 Drivers of horizontal collaboration

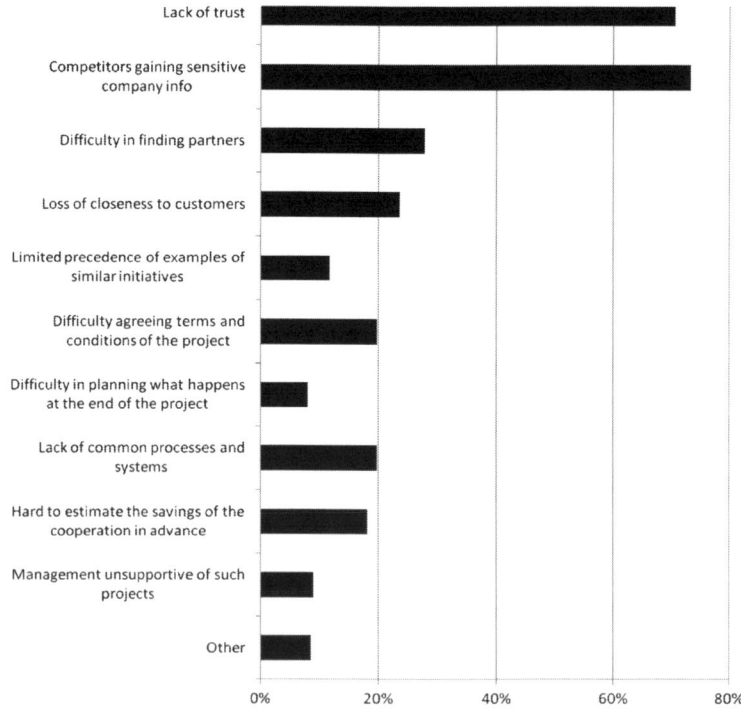

Fig. 5.5 Barriers to collaboration

5.3 Horizontal Partnerships

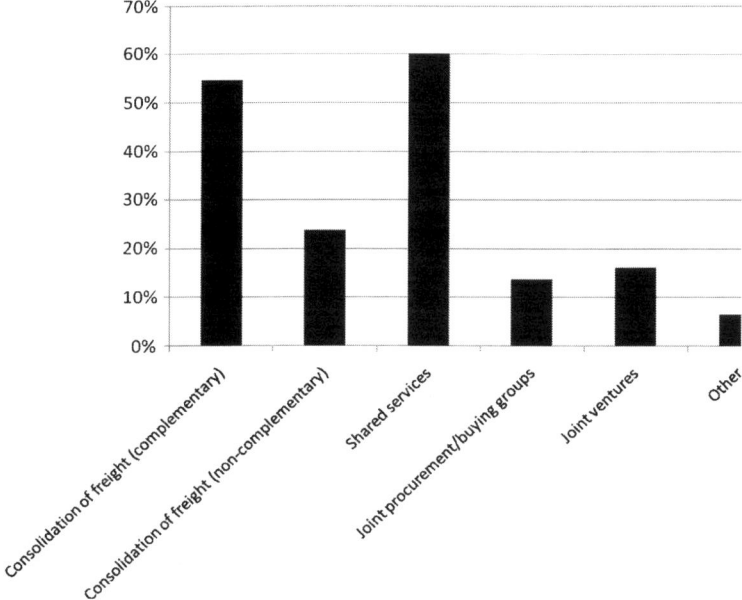

Fig. 5.6 Modes of collaboration

complementary) is conspicuous with the larger ones. These applications provide a form of collaboration that is not organic but is primarily 'co-opetitive'

Table 5.1 provides a cross tabulation of the perceived benefits and costs associated with the implementation of a formal horizontal partnership.

In summary, horizontal collaboration is a widely undertaken practice by logistics companies. It is principally used to reduce transport costs, access new markets and to enhance customer service. Shared services and freight consolidation are the most popular forms of collaboration and truckloads are the most common resource shared. Most logistics companies which practice horizontal collaboration have been in a horizontal alliance for more than five years but some have been involved in horizontal collaboration for up to 30 years. Logistics providers involved in shared services are inclined to have two to three partners whereas freight consolidators are likely to have more than three. The longer a provider has been involved in horizontal collaboration, the more partners it is likely to have. Experienced horizontal collaborators are typically involved in several different forms of collaborative alliance. The majority of relationships are asymmetric in terms of company size and the majority of providers collaborating on a local scale are small. The majority of companies are collaborating domestically but a significant percentage collaborates internationally. However, it was noted that over 50% of horizontal collaboration projects are ending prematurely. The main reasons cited for this concerned business diversification, a desire to find alternative partners and a realisation that insufficient benefits were being gained from the collaboration.

Table 5.1 Costs/benefit sharing

Are the costs shared equally?	
Yes	35%
No	65%
Are the benefits shared equally?	
Yes	48%
No	52%

5.3.2 Open Innovation

The nature of innovation is increasingly recognised as a joint or shared experience and new process and product innovations that lead to step changes in performance are being achieved through inter-organisational collaborations. This is particularly true for small and medium-sized enterprises (SMEs) which are often risk averse and have insufficient resources and expertise to innovate outside of repeating previous piecemeal and often short-term successes.

Open innovation, as defined by Chesbrough et al. (2006), is "the use of purposive inflows and outflows of knowledge to accelerate internal innovation, and expand the markets for external use of innovation, respectively..... [Open innovation] assumes that firms can and should use external ideas as well as internal ideas, and internal and external paths to market, as they look to advance their technology". The basic premise of open innovation is that the approach to the generation of new ideas is not solely undertaken internally within an organisation, but is also undertaken in a co-operative fashion with external stakeholders. Organisations should assess the opportunities to use external ideas and take advantage of the collective intelligence and know-how dispersed throughout the networks in which they operate as well as those generated internally in order to support new product and process development. A participative approach to the process of innovation redefines the boundary between a business entity and its external environment, making it more porous and part of an informal alliance of different organisations, collectively and individually working toward improving performance and exploiting new knowledge for commercial gain.

Chesbrough et al. (2006) defined open innovation in terms of the "inflows and outflows" of technologies and ideas. These "inflows" and "outflows" have been referred to as "external knowledge acquisition" and "external knowledge exploitation", respectively (Lichtenthaler 2008). The process of open innovation takes many forms. Gassmann (2006) defined five forms of open innovation: "globalisation of innovation", "outsourcing of R&D", "early supplier integration", "user innovation", and "external commercialisation". Van de Vrande et al. (2009) expressed "outflows" of knowledge (technology exploitation) as "venturing, outward licensing of intellectual property (IP) and the involvement of non R&D workers in innovation initiatives". "Purposive inflows" (technology exploration or acquisition) were classified into five practices: "customer involvement", "external networking", "external participation", "outsourcing R&D" and "inward licensing of IP" (van de Vrande et al. 2009).

Summarising these efforts, four knowledge or technology-acquisition practices—"customer-led innovation", "supplier integration", "external networking", and "inward licensing of IP", and three knowledge or technology-exploitation practices can be recognised—"venturing", "outward licensing of IP" and "the involvement of non-R&D workers in innovation initiatives".

5.3.2.1 External Knowledge Acquisition

Customer-led innovation
In order to differentiate products or services, enhance customer satisfaction or increase market share, organisations should involve their customers to contribute fresh insight and ideas (Chesbrough and Appleyard 2007; Prahalad and Ramaswamy 2000; van de Vrande et al. 2009). Businesses can benefit from conducting proactive market research with customers so that specific customers' specifications and characteristics can be communicated and exchanged between network members resulting in a product or service that better matches customers' needs (Prahalad and Ramaswamy 2000; van de Vrande et al. 2009).

Customers can also be involved in the R&D process by acting as product testers. By having customers 'consume' the product, organisations are able to improve or solve potential problems prior to the official release. This form of customer involvement is widely seen in the software industry with industry leaders such as Microsoft (Prahalad and Ramaswamy 2000).

Customers should also be encouraged to volunteer their ideas and experiences to improve quality and product variety (Chesbrough and Appleyard 2007). Such community-driven innovations are evident with YouTube (www.youtube.com), Wikipedia (www.wikipedia.org), MySpace (www.myspace.com), Linux (www.linux.co.uk), the Lego Mindstorm robot community (http://mindstorms.lego.com) and Threadless (www.threadless.com).

Supplier integration
Supplier integration is a customer-driven supply chain guideline (see Chap. 2). Within an open innovation context, supplier involvement is targeted at the new product development (NPD) process. Such a 'co-development' process refers to the sharing of technological, marketing and production information. For example, Toyota has engaged its suppliers in product development (Liker 2004) and BMW has collaborated with a small California-based high-tech supplier in developing the 'i-drive' system (Gassmann 2006).

External networking
One conspicuous element of open innovation is collaborative R&D with external solutions providers, for example, with technical service providers, engineering firms or universities. Such R&D or innovation communities (Fichter 2009; Gassmann 2006; van de Vrande et al. 2009) can be used to efficiently generate and explore new ideas. Fichter (2009) examined an innovation community centred around a joint-development project for 'coated coldset', an inexpensive base paper with a thin pigment coating which makes it whiter and easier to print on. Four

network partners—BASF, Axel Springer AG, The Flint Group and UPM Kymmene were involved.

The Nokia research centre has engaged in open innovation through collaborations with higher education institutions such as BUPT (Beijing University of Post and Telecommunications), Helsinki University of Technology, Massachusetts Institute of Technology and the University of Cambridge. By sharing resources and ideas, and accessing each partner's know-how and expertise, Nokia has been able to efficiently create, shape and disseminate innovative products such as high performance mobile platforms and new devices such as near-to-eye displays, gaze direction tracking, 3D audio and 3D video (www.research.nokia.com/openinnovation).

Another example of this practice of openness is the movie rental firm, NetFlix. NetFlix has adopted an open innovation approach by revealing problems it encounters to external organisations and offering rewards for the most appropriate solutions (Read and Robertson 2009). Not only has NetFlix acquired suitable solutions to resolve problems through this approach but it has also provided opportunities for further collaboration with the solutions providers (Read and Robertson 2009).

In addition, in order to facilitate the search for external sources of innovative ideas, new forms of mediation and 'finding solution providers' has arisen. NineSigma (www.ninesigma.com) and InnoCentive (www.innocentive.com) are examples of an open network of solution providers. The concept is that companies (or 'seekers') submit their problems to the mediators. The mediators make visible the challenges to external experts who propose their ideas.

Inward licensing of IP

In this form of open innovation, firms act as 'buyers' of external IP (intellectual property) including the licensing of patents, copyrights or trademarks. An example is the online "innovation marketplace" developed by Planet Eureka (www.planeteureka.org/marketplace/). This website is an e-market place for buying and selling innovations. Through the Planet Eureka site, a buyer can potentially find an idea to solve a problem or search for something new to market to solve others' problems. Also, sellers or inventors, patent holders, and IP owners can post descriptions of their ideas to the website. SMEs can benefit from having a 'shop window' for new ideas and inventors can be put in touch with manufacturers.

5.3.2.2 External Knowledge Exploitation

Venturing

Corporate venturing refers to the investment in, and support of, new start-up businesses. This can be to enter new industries or markets, or to develop breakthrough technologies that could lead to a step change increase in sales, profit, productivity, technical performance or quality. Parent companies may provide capital funds, experts, advice, technology, business connectivity, marketing and sales channels (Block and MacMillan 1995; van de Vrande et al. 2009).

Chesbrough (2003) discussed venturing activities in large enterprises such as Xerox, Lucent and Intel. Nokia has provided an example in forming Nokia Venture Partners for investing in technology companies (www.nokia.com). Another example concerns DSM (www.dsm.com), a life sciences company. This company specialises in creating innovative products and services primarily concerned with human and animal nutrition and health, pharmaceuticals and performance materials. DSM Venturing invests in start-up businesses with a view to gain access to new technologies and innovative products in life and materials sciences (www.dsm.com/en_US/html/venturing/home.htm).

Outward licensing of IP

Businesses can exploit their internal IP by acting as 'sellers'. In this scenario, patents become strategic assets (Gassmann 2006). Torkkeli et al. (2009) noted how large organisations such as Philips and Intel, more often exploit their non-core knowledge by licensing of IPs to other companies. Gassmann (2006) pointed out that using license and know-how transfer in 2005, IBM earned about US $1.5 billion.

Involvement of non-R&D workers

Employees are a source of new ideas which create the starting point for innovations within the organisation (van de Vrande et al. 2009). This concept of knowledge exploitation was found to be implemented by 93% of SMEs in a study by van de Vrande et al. (2009). Van Dijk and van den Ende (2002) cited several examples, including Xerox which was able to create a high degree of participation with incremental innovations, and KPN, which encouraged employees to suggest ideas in various ways: online, through suggestion boxes or on paper. These suggestions have led to substantive innovations. The value of the creative involvement of the workforce has already been emphasised in Chap. 2.

5.4 Chapter Summary

Supply chain improvement initiatives and the application of management creativity to supply chain design are often confined to squeezing out inefficiency and applying innovations to vertical structures and alliances. The adoption of sequenced supply with first-tier suppliers located on supplier parks adjacent to an OEM facility is a 'vertical' innovation found in the automotive sector. The synchronisation and zero inventory advantages created by this arrangement are founded on the proximity of partners operating in a vertical alliance. Proximate supply is analogous to a high degree of vertical integration where separate business entities work together in a closely choreographed operation rather than separate departments or units of the same organisation co-operating together. Zara, working in the textile sector, has turned vertical integration into a source of competitive advantage.

Horizontal collaboration, however, is a growing and increasingly more conspicuous part of supply chain design. In the logistics sector, horizontal alliances have become commonplace and are principally used to reduce transport costs, access new markets and to enhance customer service. Shared services and freight consolidation were found to be the most popular modes of collaboration in a survey undertaken in conjunction with the authors.

Innovations can occur from both vertical and horizontal inter-organisational collaborations. The fundamental thinking behind open innovation is that new ideas are not solely created internally within organisations, but are also created in collaboration with external partners which may be part of a vertical or horizontal network. An open approach to the process of innovation allows participating SMEs to overcome a lack of internal capabilities and competences, to share valuable resources and to reduce commercial risk.

References

Bahinipati B, Kanda A, Deshmukh S (2009) Horizontal collaboration in semiconductor manufacturing industry supply chain: an evaluation of collaboration intensity index. J Comput Ind Eng 57(3):880–895
Block Z, MacMillan IC (1995) Corporate venturing: creating new businesses within the firm. Harvard Business Press, Boston
Chesbrough HW (2003) Open innovation: the new imperative for creating and profiting from technology. Harvard Business Press, Boston
Chesbrough H, Vanhaverbeke W, West J (2006) Open innovation: researching a new paradigm. Oxford University Press, Oxford
Chesbrough HW, Appleyard MM (2007) Open innovation and strategy. Calif Manag Rev 50(1):57–76
DSM—The unlimited world of DSM. www.dsm.com. Accessed Nov 2009
DSM Venturing. www.dsm.com/en_us/html/venturing/home.htm. Accessed Nov 2009
Everington L, Lyons A, Li D (2010) Horizontal collaboration in the logistics industry. In: Whiteing T (ed) Proceedings of the LRN conference 2010: towards the sustainable supply chain, CILT (UK), Harrogate, pp 204–211
Ferdows K, Lewis MA, Machuca J (2004) Rapid-fire fulfillment. Harv Bus Rev 82(11):104–110
Fichter K (2009) Innovation communities: the role of networks of promotors in open innovation. R&D Manag 39(4):357–371
Frigant V, Lung Y (2002) Geographical proximity and supplying relationships in modular production. Int J Urban Reg Res 26(4):742–755
Gassmann O (2006) Opening up the innovation process: towards an agenda. R&D Manag 36(3):223–228
Innovation Marketplace. www.planeteureka.org/marketplace/. Accessed Nov 2009
Julka N, Srinivasan R, Karimi I (2002) Agent-based supply chain management—1: framework. Comput Chem Eng 26:1755–1769
Larsson A (2002) The development and regional significance of the automotive industry: supplier parks in Western Europe. Int J Urban Reg Res 26(4):767–784
Lichtenthaler U (2008) Open innovation in practice: an analysis of strategic approaches to technology transactions. IEEE Trans Eng Manag 55(1):148–157
Liker JK (2004) The Toyota way: 14 management principles from the world's greatest manufacturer. McGraw-Hill, New York
Lydeka Z, Adomavicius B (2007) Cooperation among the competitors in international cargo transportation sector: key factors to success. Eng Econ J 1(51):80–90

References

Naesens K, Gelders L, Pintelon L (2009) A swift response framework for measuring the strategic fit for a horizontal collaborative initiative. Int J Prod Econ 121(2):550–561

NineSigma, Inc. www.ninesigma.com. Accessed Nov 2009

Nokia Research Center. www.research.nokia.com/openinnovation. Accessed Nov 2009

Nokia: Connecting People. www.nokia.co.uk. Accessed Nov 2009

Open Innovation: Innovative Management. InnoCentive Inc. www.innocentive.com. Accessed Aug 2009

Prahalad CK, Ramaswamy V (2000) Co-opting customer competence. Harv Bus Rev 78(1):79–87

Press Release: Nokia Venture Partners Announces $100m Later-Stage Fund, December 07, 2004. www.nokia.com, www.nokia.com/press/press-releases/archive/archiveshowpressrelease?newsid=971428. Accessed Nov 2009

Read S, Robertson D (2009) Implementing an open innovation strategy: lessons from Napoleon. Strateg Dir 25(6):3–5

Torkkeli MT, Kock CJ, Salmi PAS (2009) The "open innovation" paradigm: a contingency perspective. J Ind Eng Manag 2(1):176–207

Van de Vrande V, Jong JPJD, Vanhaverbeke W, Rochemont MD (2009) Open innovation in SMEs: trends, motives and management challenges. Technovation 29(6–7):423–437

Van Dijk C, Van den Ende J (2002) Suggestion systems: transferring employee creativity into practicable ideas. R&D Manag 32(5):387–395

Chapter 6
Modelling and Simulating Supply Chains

6.1 Introduction

The dynamism of the business environments in which many organisations are operating has resulted in the acceptance that competitive advantage is often regarded as a temporary phenomenon. Organisations require lively and flexible management systems to be able to respond to market changes and retain competitiveness. Traditional organisations are structured into functional departments, utilising the concepts of the division of labour, or work specialisation. Such a configuration presents many disadvantages: it is not a customer-focused structure, it can result in sluggish performance, it does not show in a clear manner the products or services that will be provided to customers and it does not reveal how work flows satisfy the customers' requirements. Functional and individual objectives are often pursued with such structures, inadvertently ignoring the global objective of the organisation. This problem is exacerbated when the organisation grows in size and complexity. These disadvantages become even more apparent and problematic when the supply chain is taken into account. In such circumstances, the local objectives of the organisations making up a supply chain cannot easily be aligned with the overall ambitions of the entire network. Attention is increasingly focused on critical and value-adding processes due to their direct influence in business success; therefore organisations are evolving from hierarchical structures to a more integrated perspective where business process management crosses the functional boundaries of an enterprise and the organisational boundaries of a supply chain.

Enterprise modelling helps managers to understand the complex and dynamic nature of organisations and design new solutions to improve their performance. The use of simulation models to support decision-making permits the evaluation of the operating performance prior to the execution of a new design and the evaluation of the benefits derived from the consideration of alternative scenarios.

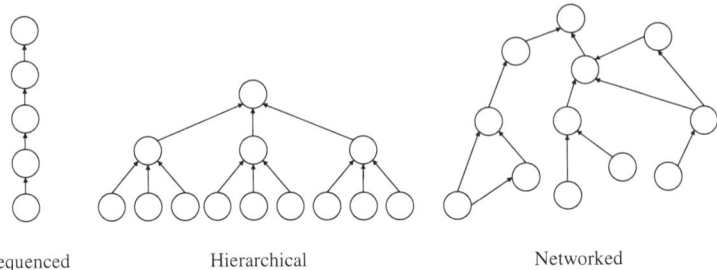

Fig. 6.1 Supply chain topologies

Effective modelling and simulating of supply chains requires the understanding of modelling frameworks and architectures which provide the means to articulate and build a precise description of the structure of the supply chain. The use of specific modelling techniques can also show interconnections between business processes and decompose them to the desired level of detail for the effective understanding of the chain.

6.2 Supply Chain Structures

Supply chains can be regarded as sets of organisations, business processes and human resources supported by information management and physical infrastructures that begin with the sourcing of raw materials and end with the delivery of products to the consumer. A supply chain can be analysed from the standpoint of its structure or from its business processes. Different topologies of supply chains exist. The three basic topologies are depicted in Fig. 6.1. These are sequenced (as a real chain), hierarchical (with a central organisation) and networked (with no dominant organisation).

Hierarchical supply chains can be found in the automotive sector and, in general, in supply chains where an original equipment manufacturer (OEM) acts as the central company. The structural aspects of a hierarchical supply chain concern at least the following:

- Supply chain members; all of the separate business entities are indirectly or directly connected to the central company through their suppliers or customers, from the point of origin to the point of consumption. Primary members carry out value-added activities, secondary members provide the resources, skills and utilities for the primary members. At the consumption point there is no more value added to the product or service.
- Structural dimensions of the network; the horizontal structure is defined by the levels of transformation of the supply chain, from raw materials to finished products. The vertical structure is defined by the number of suppliers or customers at each level.

6.2 Supply Chain Structures

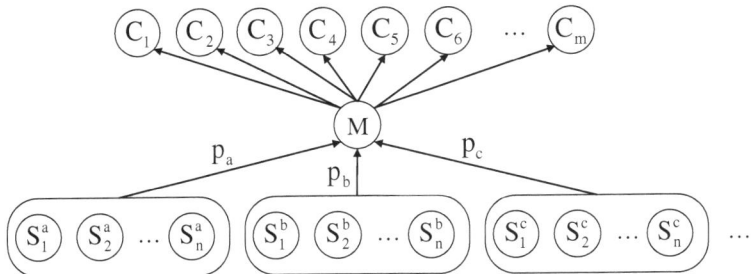

Fig. 6.2 A section of a supply chain

- Different types of nodes; a node can be any one of the business entities within the hierarchy. Depending on their importance to the central company, nodes can be 'critical', 'important' and 'normal' which are, respectively, 'managed', 'monitored' and 'unmanaged' by the central company.

Networked supply chains are more difficult to characterise as a central company does not exist, and the supply chain boundaries are hardly recognisable. Such types of supply chain can be identified by taking any node and navigating through its suppliers and customers, and so on. Figure 6.2 depicts a section of a supply chain, from the point of view of node 'M' (manufacturer). 'M' is supplied with parts 'p_a', 'p_b', 'p_c'... and it can choose from different suppliers for each part (for example, part 'p_a' can be supplied by suppliers 'S_1^a' to 'S_n^a'). Depending on the suppliers selected by 'M', the supply chain acquires different structures. On the other hand, 'M' has 'm' customers (nodes 'C_i') to sell its products. These customers can select 'M' as a supplier of their purchases. Other components of this supply chain can be observed from the point of view of each supplier 'S_j^p' or each customer 'C_i'.

6.3 Supply Chain Management

Supply chain activities can be classified according to their decision levels (strategic, tactical and operational). At the strategic level, activities concern network design, network optimisation, partnership formation, information technology investment and make-or-buy decision making. Tactical-level activities include supply contracts, demand planning, production planning and sourcing. At the operational level, activities such as production scheduling, manufacturing, distribution and inbound/outbound operations are performed.

Business processes carried out in a supply chain can be divided into intra-enterprise processes and inter-enterprise processes. The intra-enterprise business processes are those that take place within each node, while the inter-enterprise business processes affect more than one node in the chain.

Fig. 6.3 Intra-enterprise and inter-enterprise business processes

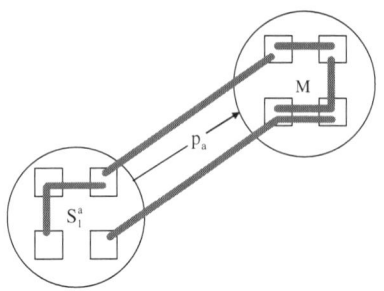

Figure 6.3 is a representation of intra-enterprise and inter-enterprise business processes in the nodes 'M' and S_1^a' of the section of the supply chain displayed in Fig. 6.2. Squares inside the nodes represent business units (departments for example). Business processes are executed inside a business unit, or between two or more business units, inside or outside the enterprise.

Davenport et al. (1990) stated that a business process always has a customer and Childe et al. (1996) added that a business process must be initiated by, and must provide results to a customer, who may be internal or external. These concepts, initially stated for intra-enterprise processes, can be extended to inter-enterprise processes in a supply chain.

Consumer satisfaction is the ultimate goal of a supply chain so it is important to try to optimise supply chain performance. Business process modelling techniques can be used to represent the processes of the supply chain. There are many techniques that can be used to model business process systems, but some of them are not appropriate to use in complex, multi-tier supply chains. Business process modelling does not only include internal organisational processes, but organisations are considered as entities carrying out processes across the supply chain. For this reason, traditional business process modelling has been extended to business network modelling.

There are some models that can be used to describe organisational and supply chain processes. The capability maturity model (CMM) is a reference model that explains the transition from the immature to the mature management of processes (Harmon 2003). CMM defines five stages that describe the processes from an immature state with an initial level of process maturity to a mature state with optimised processes. This model aims to be cognisant of the supply chain stage, to represent using diagrams the current situation in order to suggest the actions to improve this situation. The supply chain management maturity levels defined by Stemberger et al. (2005) are graphically explained in Fig. 6.4.

According to Davenport (1993), processes are defined as "simply a structured set of activities designed to produce a specified output for a particular customer or market". Hammer and Champy (1993) described a business process as "a collection of activities that takes one or more kinds of input and creates an output that is of value to the customer. A business process has a goal and is affected by events occurring in the external world or in other processes". Eriksson and Penker (2000)

6.3 Supply Chain Management

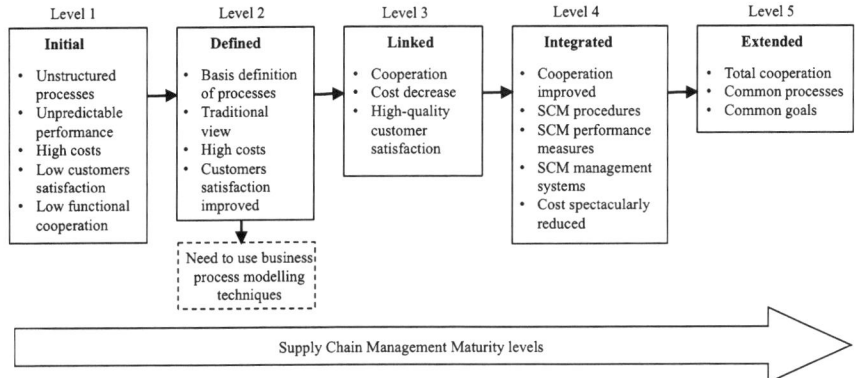

Fig. 6.4 SCM maturity levels (Stemberger et al. 2005)

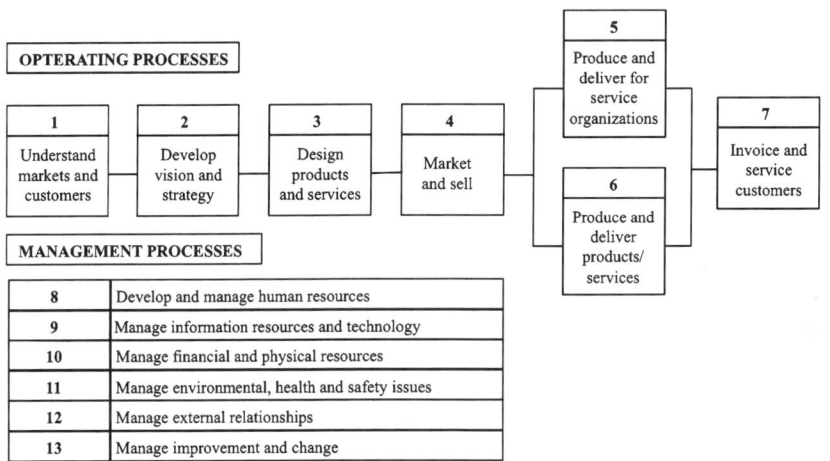

Fig. 6.5 Business process classification: operating and management processes (APQC 1996)

support Davenport's view: "a business process underlines how work is made rather than describing products or services that are the outputs of a process". Mili et al. (2003) added that "the activities of a business process are performed by actors playing particular roles, consuming some resources and producing others".

In 1996, the American Productivity and Quality Center (APQC) supported by a number of diverse international corporations, developed a process classification framework that contained two types of business process: operating processes, and management and support processes. These are represented in Fig. 6.5.

Business processes, as sets of processes and sub-processes integrated in an organisation, are difficult to understand, because they are complex and, sometimes, confused. A model is an attempt to recreate a faithful representation of reality and is used to organise and document all of the information about a system. Curtis et al. (1992) argued that there are four points of view in a business process: the functional view ('what') that represents the functional dependency between process elements; the dynamic (behavioural) view ('when' and 'how'), that provides sequencing and control information about the process; the informational view, that includes the description of the relation between the entities that are produced, consumed or otherwise manipulated by the process; and the organisational view ('who' and 'where') that describes who performs each task or function and where it is undertaken within the organisation.

Business Processes Management (BPM) is a collection of tools and techniques that is used to model, manage and optimise the business processes of an organisation. In organisations, processes are the natural way to work; processes establish a flow inside and across the boundaries of an enterprise in order to achieve, in an effective and efficient way, the objectives of organisations and the expectations and requirements of customers. BPM offers an integrated perspective, as the outputs of certain activities are the inputs of others, therefore internal co-operation can be established.

There are two complementary ways to improve business process management: continuous improvement and re-engineering. Continuous improvement is synonymous with kaizen and is primarily associated with eliminating or reducing tasks that are not value-adding from the customer point of view, or modifying activities because they are not sufficiently contributing to providing value. Plan, Do, Check, Act (PDCA) provides a systematic approach for the resolution of problems or for business process improvement. This approach tries to co-ordinate continuous improvement efforts to achieve better quality products and services, and to improve the processes that are involved. The main objective is to adapt business processes across a supply chain to new market requirements in order to foster competitiveness. The PDCA approach is often driven by a series of consecutive activities: identify the people and skills needed in the project, select and identify the problem to solve, understand the initial situation, analysis, develop a solution and action plan, present the results, review and evaluate performance and any feedback.

Creation or radical change of processes is known as re-engineering. According to Hammer and Champy (1993), re-engineering is "the fundamental rethinking and radical redesign of business processes to achieve dramatic improvements in critical, contemporary measures of performance, such as cost, quality, service and speed". A re-engineering approach is usually divided into the following stages: select people, analyse customers' and business requirements, understand the present process, analyse and generate creative and innovating ideas for the re-design of processes, design the new process, implement the new process and control the results. The relevance of re- engineering activities for a supply chain is related to the need for enterprises to adapt their business to new collaborative requirements.

6.4 Modelling Frameworks and Architectures

In the enterprise modelling domain, a framework is a fundamental structure which allows the definition and classification of the main sets of concepts to model. An architecture is a precise description of the structure of an enterprise describing the composition of systems, their relationships with the environment, and the guiding principles for the design and evolution of an enterprise.

The dominant enterprise modelling frameworks and architectures that have received most attention from researchers and practitioners include the following:

- CIMOSA Architecture
- PERA Architecture
- GIM Architecture
- ARIS Architecture
- GERAM Architecture and Methodology
- Zachman Framework
- DoDAF Architecture Framework
- TOGAF Architecture Framework

Computer Integrated Manufacturing Open System Architecture (CIMOSA) (AMICE 1993) provides a framework to analyse the evolving requirements of an enterprise and translates these requirements into a system which enables and integrates the functions that match the requirements. The original aim of CIMOSA was to elaborate an open system architecture for CIM and to define a set of concepts and rules to facilitate the building of future CIM systems. CIMOSA supports a process-based modelling approach with the goal to cover essential enterprise aspects (functional, behavioural, resource, information and organisational) in one integrated model. CIMOSA offers a comprehensive and formal set of techniques to model supply chains.

Purdue Enterprise Reference Architecture (PERA) (Williams 1994) provides the capability for modelling both the human and the technological components of an enterprise in addition to the information and control systems. PERA facilitates the modelling of an enterprise from two different perspectives, one from the external environment or business environment and the other from the internal environment, or according to company resources such as information and manufacturing technologies. The PERA diagram can be used recursively to describe the relationships between the nodes of a supply chain (Li and Williams 2000).

GRAI Integrated Methodology (GIM) (Chen et al. 1997) embodies the GRAI conceptual model, a representation of an enterprise decomposed into three subsystems: a physical system, a decision system and an information system. The GIM modelling framework has three dimensions: views, life cycle, and abstraction level. The GIM structured approach is a guide to show how to analyse and design an enterprise system and the GIM modelling formalisms (languages) concern the GRAI grid and GRAI nets for decision system modelling, IDEF0 for physical system modelling and functional system modelling and E/R for information system modelling.

Generalised Enterprise Reference Architecture and Methodology (GERAM) (Bernus 1999) encompasses all knowledge needed for enterprise engineering and integration. GERAM was developed by an IFAC/IFIP Task Force on Architectures for Enterprise Integration. The development started with the evaluation of existing architectures and frameworks for enterprise integration (mainly CIMOSA, PERA and GIM). GERAM was defined through a pragmatic approach providing a generalised framework for describing the components needed in all types of enterprise engineering processes. The GERAM aim was to facilitate the unification of methods of several disciplines used in the process of change, and included methods of industrial engineering, management science, control engineering, communication and information technologies.

Architecture of Integrated Information Systems (ARIS) was developed by Scheer (1996) at the University of Saarbruecken in Germany. The conceptual design of ARIS is based on an integration concept which is derived from a holistic analysis of business processes. ARIS allows the development of a model for business processes containing all basic features for describing them following the concept of a lifecycle model. ARIS tools ensure a consistent description from business management-related problems all the way down to their technical implementation. The ARIS architecture forms the framework for the development and optimisation of integrated information systems as well as a description of their implementation.

The Zachman Framework (Zachman 1987) is a logical structure for classifying and organising the descriptive representations of an enterprise. The Zachman matrix constitutes the intersection between the roles in the design process, i.e. 'owner', 'designer' and 'builder', and the product abstractions, i.e. 'what' (material) it is made of, 'how' (process) it works and 'where' (geometry) the components are, relative to one another. The Zachman Framework is designed to be a classification schema for organising architecture models and provides a view of the models needed for enterprise modelling.

Department of Defence Architecture Framework (DoDAF) was formerly called C4ISR and the name reflects its original target customer-group and market. C4 refers to systems for military operations, and ISR is Information System Resources. DoDAF is being further developed by the FEAC Institute of Washington DC in close co-operation with the Air Force, the Navy, the Army and Pentagon. DoDAF is particularly suited to large systems with complex integration and interoperability challenges, characteristics found in many supply chains.

The Open Group Architecture Framework (TOGAF) was developed in 1997 by the Architecture Forum of The Open Group, an international industry consortium. TOGAF is a holistic approach to modelling at four levels: business, application, data, and technology. Nowadays it is widely used in the US, UK and Japan. Most enterprise applications' vendors are members and have access to TOGAF, and to services helping them to qualify as authorised and certified TOGAF compliant providers of the methodology.

6.5 Supply Chain Modelling Techniques

Models are needed in order to understand the complex and dynamic nature of organisations and help design new supply chain arrangements. There are many business process modelling techniques but not all of them are adequate to model supply chains from the necessary perspectives (functional, dynamic, informational and organisational). This section summarises a set of techniques for supply chain modelling. These techniques are capable of displaying interconnections between processes and decompose them to the desired level of detail within the supply chain. They are also able to construct 'as is' and 'to be' views of the supply chain processes in order to depict both the existing and planned future states.

Data Flow Diagramming (DFD) is a technique for graphically depicting the flow of data, internal processing steps and data storage elements for business processes (Kettinger et al. 1997). These diagrams allow data flow and processes as well as the transformations they undergo to be modelled. DFDs do not model and take account of material flows and people (Yourdon 1989). DFDs do show the data flows between entities (for example, two different organisations within the supply chain). Each process can be divided into sub-processes in order to provide sufficient process detail. It does not facilitate information related to decisions (decision view) and process sequence (functional view) but it is a suitable approach to model the informational view within a supply chain.

Entity/Relationship diagrams (ER) are network models that describe, with a high level of abstraction, the data distribution stored in a system (Chen 1976). ER modelling is used to produce a type of conceptual schema or semantic data model of a system, often implemented in a relational database. ERs only show data and their inter-connections and do not represent the structure for modelling other elements of the process. ER diagrams are completely static representations and do not provide time-phased information (Giaglis 2001).

State Transition Diagrams (STD) are used to analyse and design systems in real time. These diagrams try to overcome the restrictions of the static nature presented by ER diagrams. STDs provide specific information about the time sequence related to the different events within the system. Their notation is very simple and consists of rectangular boxes that represent the different states and arrows that show the changes in the states (transitions).

Integrated Definition for Function Modelling (IDEF) is a family of modelling techniques developed as a system of notational formalisms to capture processes and data structures in an integrated manner. Its origin is associated with the US Defence Department which identified the need to improve its manufacturing processes. The ICAM program (Integrated Computer-Aided Manufacturing) was developed. The IDEF family is made up of many techniques, of which those related to business processes are IDEF0 and IDEF3. The full suite of techniques includes IDEF1, IDEF1X, IDEF2, IDEF4 and IDEF5.

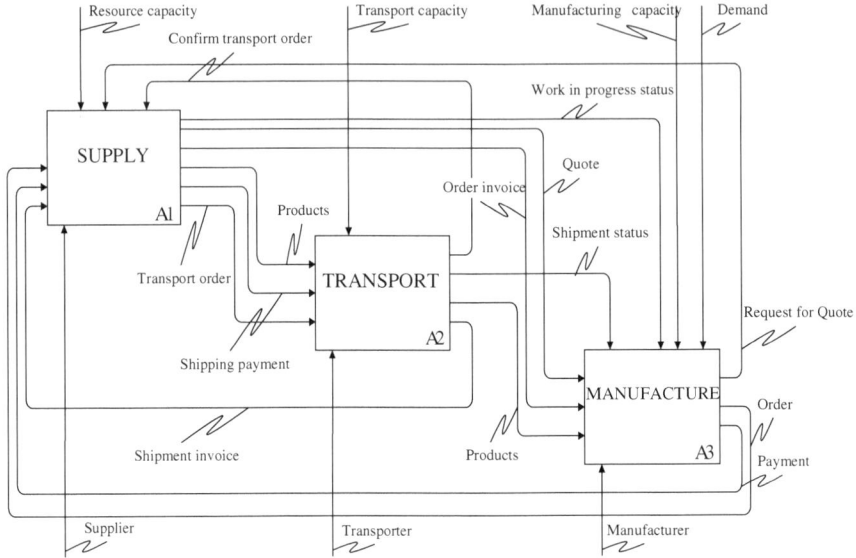

Fig. 6.6 IDEF0 model of a supply chain activity

IDEF0 was designed to model the decisions, actions and activities of an organisation or system, and as such, it is used mostly for functional modelling (Mayer et al. 1995a). It is regarded as a simple but powerful technique, widely used in industry during the analysis stage in a re-engineering effort. It allows the identification of processes and their interfaces. IDEF0 uses a single type of notation in its graphical representations known as Input-Control-Output-Mechanism (ICOM). IDEF0 allows a hierarchical or top-down decomposition approach to analyse processes at multiple levels of abstraction (Fig. 6.6). Al-Hakim (2005) used IDEF0 to map inter-organisational dependencies with the SCOR model employed to standardise processes.

IDEF3 was developed with the aim of offsetting the restrictions of IDEF0. IDEF3 describes processes as organised sequences of facts or activities. It captures temporal, precedence and causality relations between processes and events as well as modelling decisions in a form that is natural to domain experts who may be non-technical with respect to the concept of modelling (Mayer et al. 1995b). Danso-Amoako et al. (2004) used IDEF3 to support process modelling in a steel supply chain.

Role Activity Diagrams (RAD) are used to outline the activities under the responsibility of each role and the interaction between them and with external events. A role is defined as the desired behaviour of human resources within the supply chain. RAD is a very different technique because its first unit for the analysis of modelling business processes is the concept of role. RADs are, in fact, object state transition diagrams that explain how a role object changes state as a result of the actions and interactions which occur. This technique is

6.5 Supply Chain Modelling Techniques

optimum when it is used in those contexts in which human resources are a critical factor in the development of processes of the organisation and contribute relevant information about process responsibilities of each organisation within the supply chain. Murdoch and McDermid (2000) developed a modelling approach that comprises a role context chart, which depicts a top-down view of roles, and RADs, representing the timing of sequential tasks and the interactions between the roles. Each chart is placed in a frame indicating its supply chain decomposition level.

Role Interaction Diagrams (RID) identify the roles and the interactions associated with these roles in a process. A RID uses a matrix where roles are connected to activities. Although these diagrams are more complex than flow diagrams, they are very intuitive and easy in their reading. RIDs are adequate for the design of workflows, with processes that imply co-ordination of interrelated activities. Pelletier et al. (2005) used the modelling language TALMOND to represent a gas transport supply chain with an agent-based approach. With this language, it is possible to depict business situations using four types of diagrams. One of them is a RID that depicts the roles in the process.

Object-Oriented Methods (OOM) are used to model and programme a process characterised as objects, which are developed and transformed by activities throughout the process. It uses the objects as essential construction blocks and combines data structure (attributes) and functions (operations) in a single entity. There is a diversity of techniques based on Object-Oriented Methods, but among all of them, the most recognised is Unified Modelling Language (UML). UML offers a way to model conceptual objects as business processes and system functions. UML consists of nine different diagrams, each one of which explains the static or dynamic approach of the system: classes, objects, state charts, activity, sequence, collaboration, use-cases, components and deployment diagrams. Agent UML is an extension to the standard UML to accommodate the distinctive requirements of agents in a supply chain. It is language for a full lifecycle specification of agent-based systems development. Research has been undertaken on applying agent UML to a supply chain management example (Huget 2002).

The Business Process Modelling Language (BPML) is a meta-language for the modelling of business processes (BPMI 2002), just as XML is a meta-language for the modelling of business data. BPML provides an abstracted execution model for collaborative transactional business processes based on the concept of a transactional finite-state machine. BPML represents business processes supported by control flow, data flow, and event flow, while adding orthogonal design capabilities for business rules, security roles, and transaction contexts. Defined as a medium for the convergence of existing applications toward process-oriented enterprise computing, BPML offers explicit support for synchronous and asynchronous distributed transactions, and therefore can be used as an execution model for embedding existing applications within e-business processes as process components.

6.6 Supply Chain Simulation

A supply chain can be regarded as a complex system. The performance of a network design of a supply chain is difficult to predict prior to its execution. Simulation allows the evaluation of operating performance prior to the execution of a design. The use of simulation models to support decision-making is increasingly valued in supply chain design.

Supply chain simulation plays an important role among the techniques that can support a multi-decisional context mainly by facilitating complex what-if analyses evaluating benefits derived from alternative scenarios. The adequate consideration of market, logistics and production uncertainties is critical for taking successful decisions in a supply chain. Demand uncertainty, in particular, is an important factor in the supply chain design and its subsequent operations. Determining the adequate replenishment policies and stock levels in each node of a supply chain to meet a desired level of customer satisfaction, taking into account demand uncertainty, requires complex models and simulation techniques that provide the means to capture such complexity, evaluate alternatives and provide a guide to make suitable decisions.

Supply chain managers can use simulation as a decision-making support tool for addressing questions such as:

- Which suppliers are most suitable?
- Which supplier policy is achieving the best result for deliveries under a given demand pattern?
- Which supplier policy is the most robust under fluctuations in demand?
- Which inventory policy provides the greatest cost?
- What would be the impact of a given capacity increase?
- What is the balance between service level and inventory cost?
- What is the impact of the accuracy of information on manufacturing performance (cycle times, service level, etc.)?
- How can the performance of the supply chain be improved?

Thus, simulating a supply chain can evaluate the pursuit of different goals through the rigorous analysis of varying supply chain structures. Terzi and Cavalieri (2004) proposed a relevant classification schema structured into three sub-criteria: objectives, processes and morphology.

- Objectives:
  - Network supply chain design: simulation is used as a decision support system for logical modelling and industrial node configuration or for node localisation such as placing a supply chain node in a determined geographic site.
  - Supply chain strategic decision support: simulation is applied in a supply chain to evaluate strategic alternatives (for example, vendor-managed inventory versus conventional downstream replenishment).

- Processes:
  - Demand and sales planning: simulation processes can deal with stochastic demand generation.
  - Supply chain planning: simulation processes related to production planning and distribution resource allocation, under supply and capacity constraints.
  - Inventory planning: simulation processes supporting multi-inventory planning.
  - Distribution and transportation planning: simulation of distribution centres, site localisation and transport planning in terms of times and costs.
  - Production planning and scheduling: simulation processes dealing with production management of the manufacturing nodes.
- Morphology:
  - Supply chain ownership, which distinguishes two possible conditions: single ownership with nodes (manufacturers, distributors, financial sites, etc.) distributed all over the world; multi-ownership, the case of autonomous enterprises joining a logistics network for example.
  - Supply chain levels: with regards to the number of tiers within a supply chain.

Many different simulation techniques exist and, depending on the pursued goal, the most adequate has to be selected in a supply chain simulation scenario. Kleijnen and Smits (2003) distinguished four types of supply chain simulation techniques: spreadsheet simulation, system dynamics (SD), discrete-event dynamic systems (DEDS) simulation and business games (BG).

Spreadsheet simulation has a clear advantage (Powell 1997): every business computer has a spreadsheet and skills in spreadsheet use are highly desired for managers. Since the development of linear and non-linear optimisers for spreadsheets and Monte Carlo simulation add-ins, spreadsheet simulation has been widely used to support business decisions. However, spreadsheets have disadvantages: there are many sophisticated algorithms that are not well-suited to the spreadsheet platform, the spreadsheet calculation can be very slow for large models, data and calculations cannot be separated, and spreadsheets have limited flexibility for modifying or enlarging models.

Forrester (1961) developed Industrial Dynamics, later extended and called System Dynamics (SD). He developed a model for a theoretical supply chain with four links: retailer, wholesaler, distributor, and factory and examined how they react to deviations between actual and target inventories. A supply chain can be modelled trough SD as a system with six types of flows (materials, goods, personnel, money, orders, and information) and inventories (for example, WIP). Managers' decisions change the rate of some variables (for example, production rates), which change some flows and some types of inventory. Managers act taking corrective actions when undesirable deviations occur. SD is a rigorous method for qualitative description, exploration of complex systems in terms of their processes, information, strategies and organisational boundaries, facilitates the modelling and

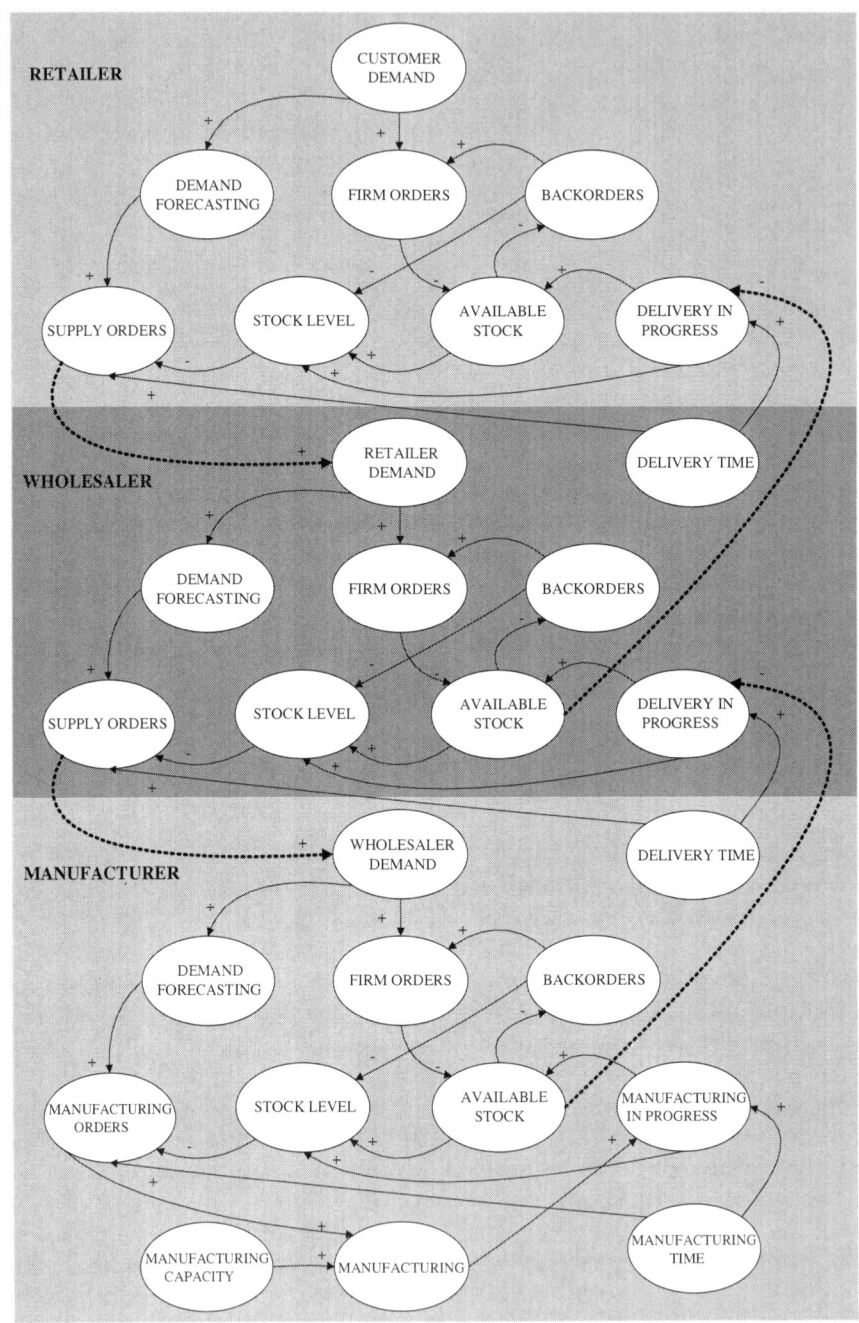

Fig. 6.7 Causal-loop diagram of a manufacturer-wholesaler-retailer supply chain

6.6 Supply Chain Simulation

analysis of the quantitative simulation for the design and control of a system structure, facilitates experimenting with business systems that require no detailed or accurate information about relationships and focuses on the dynamic behaviour of the combination of feedback loops. Causal Loop Diagrams (CLD) provide a visual representation of the feedback loops in a dynamic system and are used as the first step for an SD model (Fig. 6.7). Then inventory and flow entities are added to the diagram in order to model the accumulation and the rates of change. Finally, mathematical equations that determine the flows are introduced to reach the complete SD model.

Discrete-event dynamic system (DEDS) simulation represents individual events (for example, a customer order) and incorporates uncertainties (for example, an unreliable supplier). In DEDS simulation, the operation of a system is represented as a sequence of events. Each event occurs at an instant in time and generates a change of state in the system. The emergence of DEDS was motivated by the deficiencies of differential equation approaches to the solution of even simple business problems (for example, modelling dispatching rules with swapping between work centres).

Business games are more related to human behaviour. Managers can operate within a simulated supply chain and its environment. Kleijnen and Smits (2003) distinguished two sub-types of business games: strategic games which include several teams of players who represent companies that compete with each other in the simulated world and operational games which include a single team interacting with the simulation model either during several rounds or in real time.

From a technological point of view, complex simulation models need a specific infrastructure to be executed. Generally, supply chain simulation can be carried out according to two structural paradigms (Fujimoto 1999): local simulation when only one simulation model, executed on a single computer, is used; and parallel and distributed simulation where various models are implemented and executed on several computers in an interconnected manner. Then, a simulation model of an entire supply chain can be designed and implemented as a whole single model with all nodes, or as several models (one for each node) running in parallel mode in a single co-ordinated simulation.

A lot of SD and DEDS simulation tools exist. Specific supply chain simulation tools include i2 (Padmos et al. 1999), IBM Supply Chain Simulator (Bagchi et al. 1998), SDI Industry Pro (Siprelle et al. 1999), SCGuru (www.promodel.com), LOCOMOTIVE (Hirsch et al. 1998) and Supply Solver (Schunk 2000). General purpose simulation tools can also be used for supply chain simulation. These include WITNESS (www.lanner.com), Arena (www.arena.com), ModSim (www.modsim.org), Promodel (www.promodel.com), Vensim (www.vensim.com), AnyLogic (www.xjtek.com), Simio (www.simio.com), Dynamo, ITHINK, Powersim and Stela.

In recent years, a new paradigm for simulating supply chains has emerged. The supply chain is viewed as composed of a set of intelligent (software) agents, each responsible for one or more activities or processes and interacting with other agents. An agent is an autonomous, goal-oriented encapsulated computer programme that operates asynchronously, communicating and co-ordinating with other agents. Agents possess some special characteristics: communication for the purpose of co-operation and negotiation, learning to improve performance over time, and autonomous behaviour implying that they can act proactively within their environment rather than passively waiting for commands.

Supply chain activities can be distributed across the agents in different ways, depending on the purpose of the simulation. Fox et al. (2000) proposed the following decomposition:

- Order acquisition agent. This is responsible for acquiring orders from customers, negotiating with customers about prices, due dates, etc. and handling customer requests for modifying or cancelling their orders.
- Logistics agent. This is responsible for co-ordinating the plants, suppliers, and distribution centres in the enterprise to achieve the best possible results in terms of the goals of the supply chain (on-time delivery, cost minimisation, etc.).
- Transportation agent. This is responsible for the assignment and scheduling of transportation resources to satisfy interplant movement requests specified by the logistics agent.
- Scheduling agent. This is responsible for activity scheduling in the factory, exploring hypothetical 'what-if' scenarios for potential new orders, and generating schedules that are sent to the dispatching agent for execution.
- Resource agent. This merges the functions of inventory management and purchasing. It dynamically manages the availability of resources allowing the schedule to be executed. It is also responsible for selecting suppliers that minimise costs and maximise delivery.
- Dispatching agent. This performs the order release and real-time shop-floor control functions as directed by the scheduling agent. When deviations from the planned schedule occur, it communicates the changes to the scheduling agent. It is responsible for balancing the cost of performing the activities, the amount of time in performing the activities, and the uncertainty of the factory floor.

Multi-agent systems (MAS) provide a natural modelling framework to support collaborative supply chain decision making, simulating the effects of the decisions and policies of an entire supply chain. The agent paradigm fits very well with the real structure of a supply chain, where different entities are in charge of their own responsibilities. García-Flores and Wang (2002) described a MAS used to simulate the dynamic behaviour and support the management, over the Internet, of chemical supply chains with geographically distributed retailers, logistics, warehouses, plants and raw material suppliers modelled as an open and re-configurable network of co-operative agents. They used six classes of agents: retailers, logistics, warehouses, purchasing departments, plants and raw material suppliers, and conducted the simulation at different time scales: hours and days for scheduling decisions, months for forecasting and planning decisions, years to represent the overall performance of the supply chain.

6.6 Supply Chain Simulation

Julka et al. (2002) proposed a framework with three classes of agents:

- Emulation agents. These agents model the supply chain entities in the form of enterprise agents (representing a plant producing a set of products from a set of raw materials by utilising a set of technologies), sales agents (for modelling the sales department), warehouse agents (for modelling the materials-handling department), production agents (for modelling the operations or production department) and line agents (for modelling the physical setup converting raw materials into finished products).
- Query agents. These agents handle queries from the user and assist in supply chain analysis.
- Project agents. These agents model any new project related to the supply chain. A project is defined as a study or problem to be solved. A project agent performs the tasks needed to perform the study or solve the problem.

MAS are being used in parallel and distributed simulation due to their ability to facilitate the interaction of different companies in a supply chain, each with its own software and specific language, running in parallel, as a single, large synchronised simulation.

Some programming languages and tools for MAS include: 3APL/2APL (www.cs.uu.nl/3apl), Golog (www.cs.toronto.edu/cogrobo/main/systems), GOAL (mmi.tudelft.nl/~koen/goal.php), Jason (jason.sourceforge.net/Jason/Jason.html), Jack (www.agent-software.com.au/products/jack/), Jadex (vsis-www.inform atik.uni-hamburg.de/projects/jadex) and Jade (jade.tilab.com).

6.7 Chapter Summary

This chapter has described how supply chains can be modelled and simulated from a business process and business data perspective. Supply chains are complex and dynamic systems and modelling frameworks and architectures can provide the means to understand their components, processes and relationships in an integrated fashion. Specific modelling techniques support the analysis and design of different perspectives (functional, organisational, informational, decisional) helping managers to understand the whole system and design new solutions to improve supply chain performance. Simulation models allow the representation of complex supply chains from different perspectives. They can help assess the implications of uncertainty in processes, resources and the market, and to evaluate the benefits derived from alternative scenario testing prior to implementation.

References

Al-Hakim L (2005) Modelling Interdependencies for Electronic Supply Chain Management. IRMA International Conference, 1002–1006

American Productivity & Quality Center (1996) Process Classification Framework. APQC's International Benchmarking, Clearinghouse

AMICE (1993) CIMOSA: open systm architecture for CIM, second edition, Springer, Berlin

Bagchi S, Buckley M, Ettl M, Lin G (1998) Experience using the IBM supply chain simulator. Proc 1998 Winter Simul Conf

Bernus P (1999) GERAM: generalised enterprise reference architecture and methodology version 1.6.3.http://www.cit.griffith.edu.au/~bernus/taskforce/geram/versions/geram1-6-3/v1.6.3.html

Business Process Management Initiative (2002) Business process modeling language. Specification version 1.0, 13 November, 2002. http://www.bpmn.org/

Chen D, Vallespir B, Doumeingts G (1997) GRAI integrated methodology and its mapping onto generic enterprise reference architecture and methodology. Comput Ind 33:387–394

Chen PP (1976) The entity-relationship model—toward a unified view of data. ACM Trans Database Syst 1(1):9–36

Childe SJ, Maull R, Mills B (1996) UK experiences in business process reengineering. Business Process Resource Centre, Warwick University

Curtis B, Kellner MI, Over J (1992) Process modeling. Commun ACM 35(9):75–90

Danso-Amoako MO, O'Brien WJ, Issa RA (2004) A case study of IFC and CIS/2 support for steel supply chain processes. Proceedings 10th International Conference Computer Civil and Building Engineering (ICCCBE-10), Weimar, Germany

Davenport TH (1993) Process Innovation: Reengineering Work through Information Technology. Harvard Business School Press, Boston

Davenport TH, Short JE (1990) The new industrial engineering: information technology and business process redesign. Sloane Manage Rev 31:11–27

Eriksson HE, Penker M (2000) Business modeling with UML: business patterns at work. John Wiley and Sons, New York, USA

Forrester J (1961) Industrial Dynamics. MIT Press, Cambridge, and Wiley, Inc., New York

Fox MS, Barbuceanu M, Teigen R (2000) Agent-oriented supply-chain management. Int J Flexible Manuf Syst 12:165–188

Fujimoto R (1999) Parallel and distributed simulation. Proceedings 1999 Winter Simulation Conference, IEEE, Piscataway, NJ, pp 122–131

García-Flores R, Wang XZ (2002) A multi-agent system for chemical supply chain simulation and management support. OR Spectr 24:343–370

Giaglis GM (2001) A taxonomy of business process modeling and information systems modeling techniques. Int J Flex Manuf Syst 13(2):209–228

Hammer M, Champy J (1993) Re-engineering the Corporation: A manifesto for Business Revolution. Harper Business, NY

Harmon P (2003) Business process change: a manager's guide to improving, redesigning and automating processes. Morgan Kaufmann Publishers, San Francisco

Hirsch BE, Kuhlmann T, Schumacher TJ (1998) Logistics simulation of recycling networks. Comput Ind 24:31–38

Huget MP (2002) An application of agent UML to supply chain management. CEUR workshop proceedings

Julka N, Srinivasan R, Karimi I (2002) Agent-based supply chain management-1: framework. Comput Chem Eng 26:1755–1769

Kettinger WJ, Teng JTC, Guha S (1997) Business process change: a study of methodologies, techniques and tools. MIS Q 21(1):55–80

Kleijnen JPC, Smits MT (2003) Performance metrics in supply chain management. J Oper Res Soc 54(5):507–514

References

Li H, Williams TJ (2000) The interconnected chain of enterprises as presented by the Purdue enterprise reference architecture. Comput Ind 42(2–3):265–274

Mayer RJ, Benjamin PC, Caraway BE, Painter MK (1995a) A framework and a suite of methods for business process re-engineering, in handbook of business process change: concepts, methods and technologies, Idea Group, Pennsylvania, pp 245–290

Mayer RJ, Menzel CP, Painter MK, deWitte PS, Blinn T, Perakath B (1995b) Information integration for concurrent engineering (IICE): IDEF3 process description capture method report. Technical report, Knowledge based systems Incorporated.

Mili H, Bou JG, Lefebvre E, Tremblay G (2003) Going beyond MDA: business process modeling for software reuse. Proceedings of the workshop on legacy transformation: capturing business knowledge from legacy systems—OOPSLA'2004, October 24–28, Vancouver, Canada.

Murdoch J, McDermid JA (2000) Modelling engineering design processes with role activity diagrams. Soc Des Process Sci 4(2):45–65

Padmos J, Hubbard T, Duczmal S, Saidi S (1999) How i2 integrates simulation in supply chain optimization. Proceedings 1999 Winter Simulation Conference, 1350–1355

Pelletier CMP, de Snoo C, Maruster L (2005). Modelling the Gas Transport Supply Chain with an Agent-Based Approach. Workshop SCM-ICT

Powell SG (1997) Leading the spreadsheet revolution. OR/MS Today 24:8–10

Scheer AW (1996) ARIS-Toolset: Von Forschungs-Prototypen zum Produkt. Informatik-Spektrum 19:71–78

Schunk D (2000) Using simulation to analyze supply chains. Proceedings 2000 Winter Simulation Conference, pp 1095–1100

Siprelle AJ, Parsons D, Phelps RA (1999) SDI industry Pro: simulation for enterprise wide problem solving. Proceedings 1999 Winter Simulation Conference, pp 241–248

Stemberger MI, Jaklic J, Trkman P, Groznik A (2005) The Role of Business Process Management in a lean supply chain—two case studies. Informing Science and Information Technology

Terzi S, Cavalieri S (2004) Simulation in the supply chain context: a survey. Comput Ind 53(1):3–16

Williams TJ (1994) The Purdue enterprise reference architecture. Comput Ind 24(2–3):141–158

Yourdon E (1989) Modern structured analysis. Yourdon Press Upper Saddle River, NJ

Zachman JA (1987) A framework for information systems architecture. IBM Systems Journal, 26, 3, IBM Publication G321–5298

Chapter 7
Supply Chain Performance Measurement

7.1 Introduction

The growing importance of the management of supply chains has motivated researchers and practitioners to develop and implement measures that can be used to establish supply chain performance. The measurement of supply chain performance requires the creation of an inter-organisational assessment system. Such systems can feasibly be used to identify opportunities for improved supply chain efficiency and competitiveness, to help understand how companies operating in supply chains affect each other's performance, to support the supply chain in satisfying consumer requirements and to assess the result of an implemented initiative.

However, most supply chain performance measurement systems focus on a specific supply chain dimension such as logistics, or are dyadic—too often, companies focus only on their immediate first-tier suppliers, customers, or third-party logistics providers, and as a consequence often fail to understand the implications of their decisions on the performance of those suppliers residing further upstream or downstream within the chain and, therefore, are ignorant of the performance of the whole system. Such partial or dyadic measurement systems compromise the customer-driven ideal as the supply chain is perceived as a series of disparate elements rather than an inclusive, integrated whole.

7.2 Measures of Supply Chain Performance

Any business may be able to identify a multitude of measures to provide some perspective on supply chain performance. In order to try to establish a manageable set of measures, Gunasekaran et al. (2001) proposed a key measurement-set. This concerned seven key measures regarded as 'strategic':

- total cash flow time,
- rate of return on investment,

- flexibility to meet particular customer needs,
- delivery lead time,
- total cycle time,
- level and degree of buyer–supplier partnership, and
- customer query time.

Three 'tactical' measures were proposed:

- extent of co-operation to improve quality,
- total transportation cost, and
- truthfulness of demand predictability/forecasting methods.

Four 'operational' measures were also proposed:

- manufacturing cost,
- capacity utilisation,
- information carrying cost, and
- inventory carrying cost.

Attempts to articulate a pertinent measurement framework is important in order to strive for a point of reference and provide a starting point for the specific measurement efforts of supply chain partners. Gunasekaran et al. (2004) developed a supply chain performance framework consisting of four supply chain processes: plan, source, make/assemble and deliver. Each of these processes with the exception of 'source' included measures once again classified as strategic, tactical and operational. The measures ranged from the level of perceived customer value of a product, customer query time and order entry methods to the efficiency of the purchase order cycle time. Caution was recommended in order to ensure that differences in supply chain operations are taken into account when designing and adopting a performance measurement system. In addition, the researchers suggested that it may be desirable for supply chain partners to include further measures that reflect their unique needs.

The development of customer-driven, supply chain performance measurement systems is a demanding task. Capturing cross-supply chain performance is complex and necessarily requires some degree of supply chain-wide information sharing. The following list of attributes is recommended in order for a measurement system to be customer-driven:

- strategic focus—the measurement system should be aligned with the customer-driven concept and facilitate improvement from an holistic, cross-supply chain perspective,
- a balanced approach—single metrics are simple to apply but the measurement system should account for different performance perspectives that fit into an integrated whole,
- a manageable quantity of metrics—the measurement system should focus on relatively few, insightful measures,

7.2 Measures of Supply Chain Performance

Fig. 7.1 Supply chain performance measurement system

Behavioural Measures	*Responsiveness Measures*
Bullwhip index Synchronisation index	Demand forecast MAD Inventory level Supply chain cycle time Value-adding contribution
Reliability Measures	*Cost Measures*
Stockouts Backorders	Transportation cost Inventory holding cost

- metrics' compatibility—metrics should be understandable and workable across organisational boundaries, and be complementary in establishing the performance of multi-tier chains.

The supply chain performance measurement system presented in this chapter attempts to satisfy the previous four attributes and is composed of four dimensions: 'behaviour', 'responsiveness', 'reliability' and 'cost'. The proposed system, the 'supply chain scorecard', is based on research work undertaken in conjunction with the authors by Coleman et al. (2004). The scorecard approach to supply chain performance measurement is inspired by Kaplan and Norton's (1992 and 1996) balanced scorecard, and utilised some of the measures suggested by Gunasekaran et al. (2001). Figure 7.1 illustrates the four dimensions of the measures comprising the supply chain performance scorecard.

7.3 Behavioural Measures

The magnitude of the bullwhip effect (demand amplification) is regarded as a behavioural supply chain measure. (Chap. 3 provides an explanation of the bullwhip phenomenon.) The nature of the bullwhip effect along the supply chain can be appreciated via Fig. 7.2 which represents a three-tier linear segment of an automotive supply chain. The values plotted in Fig. 7.2 represent 21 working days and illustrate high levels of amplification upstream in the supply chain.

7.3.1 Measuring the Bullwhip Effect

Bullwhip refers to the increasing variability of demand further upstream in a supply chain. A bullwhip metric necessarily establishes the amplification between tiers by comparing the variability of the demand signal from the downstream

Fig. 7.2 Illustration of demand amplification

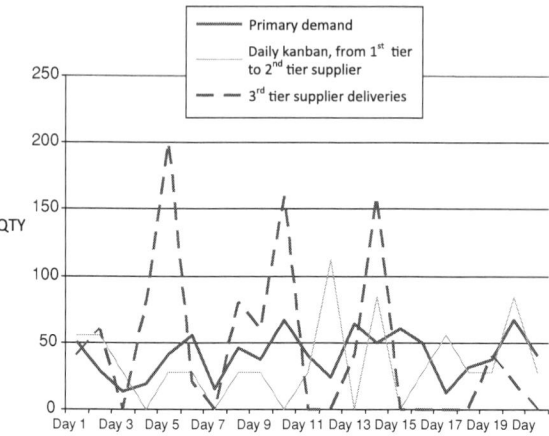

partner with the variability upstream. In Fig. 7.2, actual demand can be compared to the third tier. Variability is measured by the standard deviation of demand relative to mean demand. Equation 7.1 illustrates the bullwhip measure.

$$Bullwhip - measure = \frac{\frac{\sigma}{\mu} upstream}{\frac{\sigma}{\mu} downstream} \quad (7.1)$$

where,

σ = the standard deviation of the demand pattern, and
μ = the mean of demand pattern.

'Upstream' refers to demand/orders exhibited at first, second or third-tier suppliers, and 'downstream' refers to demand/orders at the point of origin (usually an OEM or retailer). Background on the use of this measure can be found in Coleman et al. (2004).

Another expression for the bullwhip effect has been proposed by Fransoo and Wouters (2000):

$$\omega = \frac{C_{out}}{C_{in}} \quad (7.2)$$

where,
ω refers to the magnitude of the bullwhip effect, C_{out} corresponds to Eq. 7.3,

$$C_{out} = \frac{\sigma(D_{out}(t, t+T))}{\mu(D_{out}(t, t+T))} \quad (7.3)$$

and C_{in} corresponds to Eq. 7.4,

$$C_{in} = \frac{\sigma(D_{in}(t, t+T))}{\mu(D_{in}(t, t+T))} \quad (7.4)$$

7.3 Behavioural Measures

Fig. 7.3 Illustration of the bullwhip effect behaviour

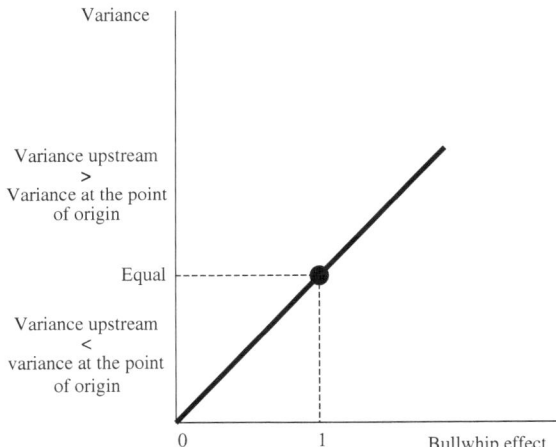

Table 7.1 Deliveries—bullwhip effect scenarios

Sales/Demand pattern	Second tier	Bullwhip index
Daily	Weekly deliveries	Typically high ($\gg 1$)
Daily	Daily deliveries	Typically low (close to 1)

The expressions $D_{out}(t,t+T)$ in Eq. 7.3 and $D_{in}(t,t+T)$ in Eq. 7.4 refer to demand during the time interval $(t,t+T)$. A bullwhip index > 1 implies that the variance of the demand registered at the upstream tier is higher than that registered at the point of origin. Situations where a bullwhip index > 1 can be found include a regular but more infrequent weekly delivery of items from upstream tiers than consumption of those items at a downstream tier. Figure 7.3 provides an illustration of the impact of demand variance against 'bullwhip'. A bullwhip index < 1 implies that the variance of the demand registered at the upstream tier is lower than that registered at the point of origin. Situations where a bullwhip index < 1 may occur include more frequent deliveries of items from upstream tiers than consumption of those items at a downstream tier. A bullwhip index equal to 1 means a perfectly balanced supply chain with no demand amplification, that is, C_{out} and C_{in} are equal.

Based on the illustration of the bullwhip effect index presented in Fig. 7.3 it is possible to suggest that even if customer demand is constant for a period examined, the size and frequency of suppliers' deliveries can have a negative impact on the supply chain index. Weekly deliveries of large batches will often result in a bullwhip index well above one. Sequenced, synchronised deliveries (one-piece flow) will result in a bullwhip index close to or equal to one.

Table 7.1 provides a simple illustration of the impact of deliveries on the bullwhip index.

7.3.2 Establishing a Synchronisation Index

The second behavioural measure concerns the notion of synchronicity of supply chain activity. A 100% synchronisation index is achieved in a supply chain if both second and third-tier suppliers produce and deliver to actual demand, but offset by the appropriate supply chain lead times. Equation 7.5 illustrates the synchronisation index for a second-tier supplier. There is an offset time of one day between the OEM and the second-tier supplier. In this case, there is no offset time between the OEM and the first-tier supplier.

$$Sync = 100 - \left(\frac{\left(\frac{\sum_{i=1}^{p} |a_i - b_{i-1}|}{p} \right)}{\mu} \right) * 100 \qquad (7.5)$$

where,

a = requested quantities (daily demand),
b = supplier deliveries (daily),
p = days for the period examined,
μ = mean demand for the period examined.

The measure calculates the mean absolute deviation (MAD) over the period examined, between the appropriate offset OEM demand and the actual demand. Synchronisation for a third-tier supplier is presented in Eq. 7.6. There is an offset time of two days between the OEM and the third-tier supplier, one offset day between the first tier and the second tier and one offset day between the second tier and the third tier. There is no offset time between the OEM and first tier.

$$Sync = 100 - \left(\frac{\left(\frac{\sum_{i=1}^{p} |a_i - b_{i-2}|}{p} \right)}{\mu} \right) * 100 \qquad (7.6)$$

where,

a = requested quantities (daily demand),
b = supplier deliveries (daily),
p = days for the period examined,
μ = mean demand for the period examined.

7.3 Behavioural Measures

Fig. 7.4 Illustration of the synchronisation index

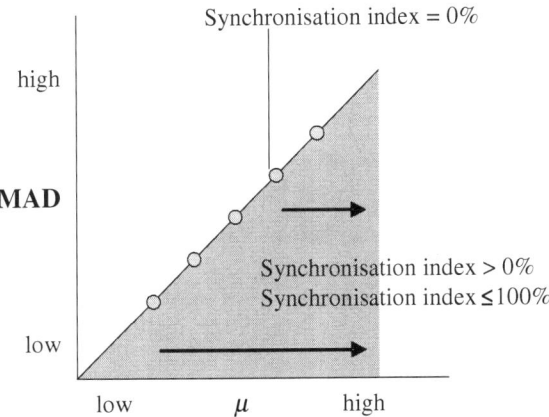

The maximum synchronisation index is 100%. The synchronisation index is directly dependent on the sum of absolute errors. A synchronisation index is close to 100% if the total sum of absolute errors divided by the number of days comprising the period of study is significantly smaller than the mean demand for the period ($MAD_{offset} < \mu$). The shaded area in Fig. 7.4 depicts a synchronisation index ranging from $> 0\%$ to $\leq 100\%$. The 45° slope represents the frontier where $MAD_{offset} = \mu$ implying a synchronisation index of 0%.

7.4 Responsiveness Measures

The suggested responsiveness measures for the scorecard tool that fit the customer-driven concept concern forecast accuracy, supply chain cycle time, pipeline inventory and value-adding contribution.

7.4.1 Forecast Accuracy

Forecast accuracy can be established in a number of different ways. One approach is to determine the mean absolute deviation (MAD) between forecasted and actual requirements. The calculation of the forecast accuracy index is presented in Eq. 7.7.

$$Forecast\_acc = 100 - \left(\frac{\left(\frac{\sum_{i=1}^{p} |a_i - b_i|}{p} \right)}{\mu} \right) * 100 \qquad (7.7)$$

where,

a = final product call-off/consumption (daily demand),
b = supplier deliveries (daily),
p = days for the period examined,
μ = mean demand for the period examined.

The resulting values of the forecast index can be interpreted in a similar manner to those of the synchronisation index. A forecast index close to 100 implies the total sum of absolute errors divided by the number of days comprising the period of study is significantly smaller than the mean demand for the period examined ($MAD_{forecast} < \mu$). A forecast index is zero if the total sum of absolute errors divided by the number of days comprising the period of study is equal or higher than the mean demand for the period examined ($MAD_{forecast} \geq \mu$).

7.4.2 Supply Chain Cycle Time

Time is the critical component of any assessment of responsiveness. Good responsiveness requires the lead time for materials to travel through the supply chain pipeline to be kept as short as possible. Figure 7.5 provides an illustration of the composition of supply chain cycle time across several supply chain tiers. At each tier, the cycle time (sometimes referred to as dock-to-dock time) includes the process time, waiting time and inventory storage time within each organisational boundary. Delivery time is included as part of the overall supply chain cycle time. The overall supply chain cycle time indicates the total length of the supply chain pipeline. A supply chain with short dock-to-dock times can usually respond to demand or market changes quickly because the time for the appropriate modifications to take effect throughout the supply chain and the restoration of an equilibrium position is shorter.

7.4.3 Pipeline Inventory

Inventory exists in different forms throughout the supply chain pipeline—raw materials, work-in-progress, finished goods, goods in transit and spare (or service) parts. An effective inventory management system maintains a low level of inventory while having sufficient to satisfy customer orders or demand. Raw materials, component parts and finished goods are taken into consideration in the scorecard. The inventory level is calculated from the ratio of the average daily inventory quantity to average daily usage quantity. Hence, the inventory level is presented in a 'days of inventory' format. The mathematical formula is represented by Eq. 7.8.

7.4 Responsiveness Measures

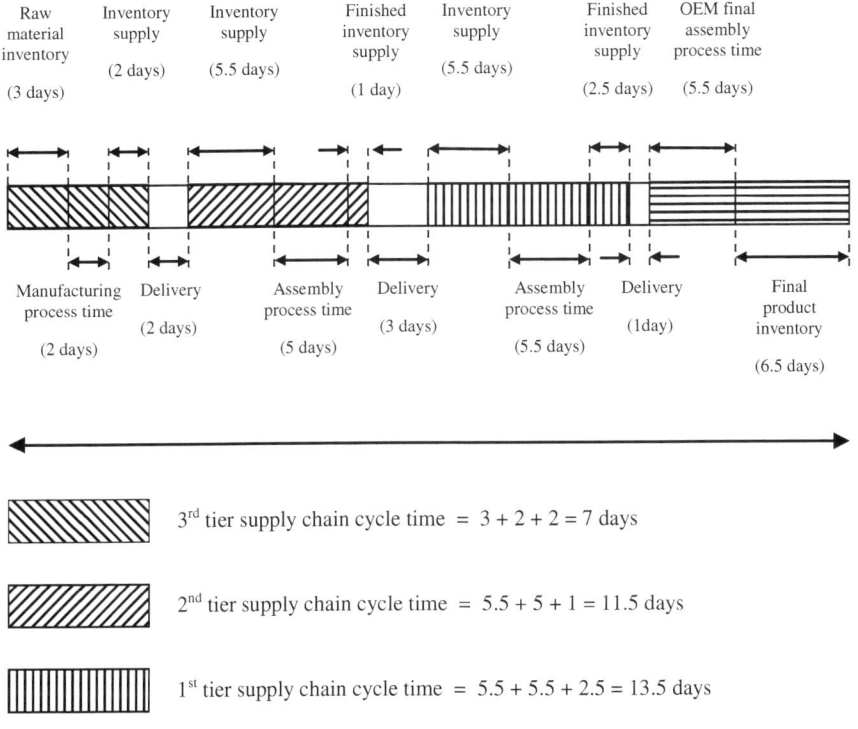

Fig. 7.5 An illustration of supply chain cycle time

$$Inventory \ Level \ = \ \frac{\left(\frac{\sum_{i=1}^{n} I_i}{n}\right)}{\left(\frac{\sum_{i=1}^{n} U_i}{n}\right)} \quad (7.8)$$

which can be simplified to

$$Inventory \ Level \ = \ \frac{\sum_{i=1}^{n} I_i}{\sum_{i=1}^{n} U_i} \quad (7.9)$$

where,

I = daily inventory quantity,
U = daily usage quantity,
n = sample size.

7.4.4 Value-Adding Contribution

Value is added to material as it moves through a supply chain until the final saleable item is produced and ultimately consumed. One of the customer-driven guidelines is to eliminate waste. Waste, or non-value adding activity, in a supply chain is an inhibitor to responsiveness. In order to monitor the proportion of value-added contribution (VAC) in a chain, the percentage of value-adding activity is expressed as follows:

$$V.A.C. = \frac{\mathrm{Process\ Time}}{\mathrm{Supply\_chain\_cycle\ Time}} \times 100 \qquad (7.10)$$

This formula provides the value-added contribution for each supply chain tier. The overall supply chain V.A.C. is derived by dividing the sum of the process times (from the involved supply chain tiers) by the overall supply chain cycle time. A low V.A.C. indicates a high level of non-value-adding activities and vice versa. The higher the V.A.C., the less non-value-adding activities within the supply chain, hence the higher the expectation of a better degree of responsiveness.

7.5 Reliability Measures

Supply chain reliability is associated with the satisfaction of customer service and is analogous to product reliability in that it conveys the ability to function without disruption and according to an agreement or specification. An unreliable supply chain is likely to have a negative impact on customer service, the partners' profitability and product quality (Waller, 2003). There are many constructs that can be used to assess supply chain reliability. These include metrics for delivery performance, order fulfillment and the proportion of schedules changed within a supplier's lead time (Stewart, 1995). However, to keep the proposed supply chain scorecard to a manageable size, only two reliability metrics are included: the stockout level and the backorder level. In several industries, for example, a late delivery will incur extra costs to both the supplier and buyer, in the form of compensation cost and production disruption cost. One way to support the attainment of a reliable supply chain is to engender trust, share appropriate data and information, and achieve a continuous and smooth material flow throughout the supply chain pipeline. Reliability assessment via the scorecard focuses on monitoring the frequency of stockout incidents and the magnitude of backorders.

7.5.1 Stockouts

The stockout level indicates the inability of an organisation to meet demand at the required time (Waller, 2003). The stockout metric in the scorecard refers to the frequency of stockout incidents. If one facility in a supply chain fails to produce according to schedule, it affects the material availability of the next supply chain facility. This can create a knock-on effect throughout the whole supply chain and incur additional direct and indirect costs.

The measurement of the stockout level in the scorecard assesses the disruption of material availability by examining the average number of days on which a stockout incident occurred per month. A high stockout level indicates many disruptions have occurred to material provision and to the desired level of customer service. It may also provide an indication of the reliability of upstream suppliers and their materials management effectiveness.

7.5.2 Backorders

In addition to measuring the frequency of disruption in material availability, the second metric in the scorecard that helps to gauge supply chain reliability is the backorder level. Ordinarily, backorder is referred to as the portion of orders that are not delivered on time. However, in this scorecard, the backorder level measures the magnitude of material availability disruption. It calculates the average quantities of materials that have been delayed in stockout incidents which were sampled in the stockout level metric.

7.6 Cost Measures

Maintaining a handle on supply chain costs is an essential element of any supply chain performance measurement system and is particularly relevant to the customer-driven concept in order to establish a supply chain cost profile and assess the evenness of the cost distribution across a chain. Two major categories of costs affecting the supply chain are introduced in this section.

7.6.1 The Cost of Holding Inventory

The cost of capital (sometimes referred to as the cost of finance or the opportunity cost), insurance, inventory obsolescence, warehousing and materials handling, deterioration, damage and pilferage, and administration and various other

Fig. 7.6 Inventory holding costs

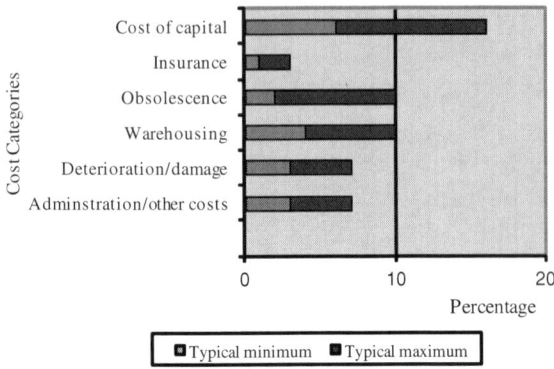

miscellaneous expenses can all contribute to inventory holding costs and be interpreted as a percentage of the value of the inventory per holding period. A general rule-of-thumb of 25% of inventory value is regularly attributed to the cost of holding inventory but costs can differ significantly from business to business and from inventory item to inventory item (Lamarre, 2003). Figure 7.6 illustrates typical minimum and maximum percentages of inventory values for the different categories of cost. The cost of capital to finance the inventory is the most significant contributor to overall holding costs. Obsolescence is an example of a cost category that can vary considerably from 10% or more for an inventory item in a high-clockspeed electronics supply chain to as low as 2% for an inventory item in a less ephemeral, slower-moving chemicals supply chain.

7.6.2 Transportation Costs

Transportation cost is a major component of the cost of logistics. Global markets and the trend to also source globally and stretch supply chains has made the role of logistics and transportation and their associated costs more conspicuous. In dynamic industry sectors where transportation is multi-country, transportation costs can differ from country to country even within the same economic block such as the European Union. For example, haulage rates in Spain are often calculated differently to haulage rates in the UK.

An example scorecard taken from an analysis of an automotive supply chain is depicted in table 7.2 for behaviour and responsiveness measures and in table 7.3 for the corresponding reliability and cost measures. The scorecards are also presented in a format that allows the presentation of the performance profiles of three alternative modes (Plan A, Plan B and Plan C) of working to the incumbent supply chain design.

7.7 Value Stream Mapping Measures

Table 7.2 Behaviour and responsiveness scorecard

Supply Chain Scorecard Part I

	Current	Plan A	Plan B	Plan C
Supply Chain Behaviour Measures				
Synchronisation index—overall (%)	35.70	260.19	134.48	130.71
First tier	99.00	99.00	99.00	99.00
Second tier	2.39	80.59	49.02	47.86
Third tier	0	0	0	0
Bullwhip measure (OEM—Tier 3)	3.31	1.04	2.25	2.25
Responsiveness Measures				
Forecast accuracy (%)				
First tier	0	17.16	0	100.00
Supply chain cycle times—overall (days)	18.66	14.11	13.54	13.19
First tier	11.20	8.52	7.33	7.50
Second tier	4.61	3.74	3.85	3.59
Third tier	2.85	1.85	2.35	2.10
Pipeline Inventory—overall (days of inventory)	12.58	8.01	7.45	7.09
First tier component inventory	5.58	2.89	1.70	1.87
Second tier finished goods inventory	2.50	1.50	2.00	1.75
Second tier component inventory	2.00	2.13	1.74	1.73
Third tier finished goods inventory	2.50	1.50	2.00	1.75
Value adding\Non—value adding (%)	48.30	76.06	81.83	85.93

In enterprise value stream mapping and supply chain value stream mapping, process measures are displayed in dedicated process boxes such as the one depicted by Fig. 7.7.

A characteristic of the value stream mapping approach (see Sect. 2.2 in Chap. 2) is that it links the flow of material and the supply chain operations with specific measures of performance and points of information in the VSM process boxes. Some of the metrics included in the process boxes used in value stream and supply chain value stream mapping are straight forward to calculate. For example, cycle time concerns the time associated with a specific process. Common VSM measures include cycle time, change over time, availability, uptime, quality rate and overall equipment effectiveness (OEE). Figure 7.8 provides an example of a completed VSM process box.

Other measures that are regularly included as entries in a VSM process box include 'takt time' and 'pitch'. Producing to takt time supports the alignment of production with demand and reveals opportunities to reduce non-value-adding activities throughout the supply chain. The specific formula used to calculate takt time is presented in Eq. 7.11.

Table 7.3 Reliability and cost scorecard

Supply Chain Scorecard—Part II				
	Current	Plan A	Plan B	Plan C
Reliability Measures				
Stockout incidents—Overall	1	0	0	0
First tier component days stockout	0	0	0	0
Second tier component days stockout	1	0	0	0
Backorders—Overall	1	0	0	0
First tier component backorders	0	0	0	0
Second tier component backorders	1	0	0	0
Cost Measures				
Transport	€ 170.91	€ 170.91	€ 170.05	€ 170.05
First tier	€ 2.47	€ 2.47	€ 2.47	€ 2.47
Second tier	€ 2.29	€ 2.29	€ 2.29	€ 2.29
Third tier	€ 166.15	€ 166.15	€ 165.29	€ 165.29
Inventory	€ 66.56	€ 8.80	€ 11.57	€ 11.13
Interest on capital cost	€ 4.06	€ 0.54	€ 0.71	€ 0.68
Warehousing	€ 6.70	€ 0.89	€ 1.16	€ 1.12
Obsolescence and depreciation	€ 50.73	€ 6.71	€ 8.82	€ 8.48
Other finance costs	€ 5.07	€ 0.67	€ 0.88	€ 0.85
Inventory saving	–	€ 693.09	€ 659.94	€ 665.15
Average cost/benefit	–	€ 693.09	€ 670.31	€ 675.52
Implementation cost	–	240.00	€ 120.14	€ 120.14
Return on investment	–	2.8879	5.5793	5.6226
Payback period (years)	–	0.3463	0.1792	0.1779

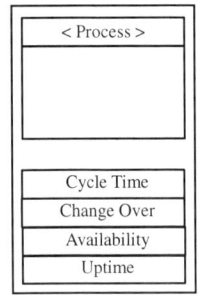

Fig. 7.7 A value stream mapping process box

$$Takt\_time = \frac{Time\_available\_per\_period}{Demand\_per\_period} \qquad (7.11)$$

Available time, the numerator in Eq. 7.11, excludes personal break time when a process has stopped and times associated with stoppages or meetings at the start and end of a shift period.

7.7 Value Stream Mapping Measures

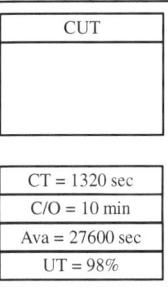

Fig. 7.8 A completed VSM process box

Pitch time uses the same principles as take time. Pitch time is used when customers do not place single orders, but rather place an order for a standard pack quantity. Pitch is the length of time required for an upstream operation or supplier to release a pre-determined pack quantity to a downstream operation or customer. The representation of pitch time is shown in Eq. 7.12.

$$Pitch\_time = takt\ time \times pack\ quantity \qquad (7.12)$$

7.8 Chapter Summary

This chapter introduced the subject of supply chain performance measurement and proposed a series of inter-organisational metrics that are compatible with the notion of customer-driven supply chain design. Despite there being many different approaches to measure supply chain performance, there is no one perfect measurement approach that can suit all supply chains. In order to guide the development of a customer-driven measurement system, the approach adopted should have a strategic focus across supply chain tiers, it should be balanced accounting for different perspectives, it should consist of a manageable quantity of metrics and the metrics should be compatible and understandable across organisational boundaries.

Four categories of measures were considered: behavioural, responsiveness, reliability and cost. Behavioural measures concern supply chain co-ordination and include the measurement of bullwhip and synchronisation. Responsiveness measures are associated with the flexibility of the pipeline of activities that constitute the supply chain and include forecast accuracy, supply chain cycle time, pipeline inventory and value-adding contribution. Reliability measures are concerned with the satisfaction of customer service and any disruptive effects of activities within the chain. Stockouts and backorders are relevant measures of reliability. Cost measures include transportation and inventory holding cost, and can be used to establish a quantitative cost profile across a chain.

References

Coleman J, Lyons A, Kehoe D (2004) The glass pipeline: increasing supply chain synchronisation through information transparency. Int J Technol Manag 28(2):172–190

Fransoo J, Wouters M (2000) Measuring the bullwhip effect in the supply chain. Supply chain manag: Int J 5(2):78–89

Gunasekaran A, Patel C, Tirtiroglu E (2001) Performance measures and metrics in a supply chain environment. Int J Operations Prod Manag 21(1/2):71–87

Gunasekaran A, Patel C, McGaughey RE (2004) A framework for supply chain performance measurement. Int J Prod Econ 87(3):333–347

Kaplan RS, Norton DP (1992) The balanced scoreboard—measures that drive performance. Harvard Business Review, January-February, 71–79

Kaplan RS, Norton DP (1996) The Balanced Scorecard. HBS Press, Boston

Lamarre R (2003) Determining the cost of carrying inventory or the magic number. Available http://www.gcrl.ca/english/magicnume.htm

Stewart G (1995) Supply chain performance benchmarking study reveals keys to supply chain excellence. Logist Inf Manag 8(2):38–44

Waller DL (2003) Operations management: a supply chain approach, 2nd edn. Thomson Business Press, London

Chapter 8
Designing Customer-Driven Supply Chains: Illustrative Cases

8.1 Introduction

The cases presented in this chapter are based on the study, analysis, simulation and prototyping of a customer-driven supply chain design in three automotive supply chains. The cases are extended versions of those originally presented in Lyons et al. (2005) and attempt to explore the principles of the 'customer-driven' supply chain (as introduced in Chapter 2)—planning and activity synchronisation through information sharing and customer-driven processes, and to do so within an empirical context.

Organisations working in the automotive sector are accustomed to operating in a highly competitive, turbulent environment. In recent years, the industry has been confronted with changeable market conditions around the globe, overcapacity in many production sites, rising costs of raw materials, increases in oil prices and the introduction of new, environmental legislation. The difficulties experienced by some in the sector have led to a number of high-profile mergers, takeovers and de-mergers. Some automakers have entered into horizontal collaborations in the form of global alliances in order to create synergies that have allowed them to share vehicle platforms and reduce product development costs. The challenges and intense competition faced by those operating in the industry have also provided the motivation to develop and adopt, and in many cases, pioneer a variety of supply chain, manufacturing and business improvement initiatives such as total quality management, just-in-time (JIT) and lean thinking.

The role of the supply chain in the automotive sector has been central to many of the changes that have been experienced. One initiative that has gained maturity is the supplier park concept of locating clusters of suppliers in close proximity to the site of final product assembly (Miemczyk et al. 2004). According to Miemczyk et al. (2004), a supplier park is the concentration of dedicated production, assembly, sequencing or warehousing facilities managed by suppliers or a third party in close proximity to one vehicle assembly facility (original equipment

Fig. 8.1 Supply chain arrangement investigated

manufacturer (OEM)). In the automotive industry, supplier parks have been adopted in several production sites across Europe. These proximate supply arrangements are characterised by outsourcing of the assembly of key modules or components to first-tier partners which assemble and deliver the modules and components in the same sequence as the final assembly of the vehicles at the OEM plant. The first-tier (and occasionally second-tier) suppliers are located on supplier parks in close proximity to the vehicle assembly plant. GM, Ford and Volkswagen are among the principal OEMs that have developed supplier parks adjacent to their European vehicle assembly plants.

Modern automotive vehicles are sophisticated, highly-engineered products made up of thousands of different modules and components. A seat set was chosen as an appropriate value stream for each of the studies as seats are indicative of a typical vehicle sub-system. They are complex, high-variety modules assembled by large first-tier suppliers. Seats sets are scheduled as single items and seat assemblers have their own deep supply chains with first, second and third-tier suppliers providing potential opportunities to explore different information sharing schemas. Figure 8.1 depicts the traditional flow of information and material across a seat value stream. It can be seen that information created by the OEM is made available only to its immediate, contiguous, first-tier supplier. Information created by the first-tier supplier is available only to the second tier and so on. Information flow is represented by the thin dotted line moving upstream through the chain. The flow of material is represented by the thick line moving downstream through the chain. This chapter describes how the arrangement can be improved and become more customer-driven if the OEM works collaboratively with its extended supply chain to bring about enhanced synchronisation within the operations of the lower tiers. It suggests a development to the existing supply chain design, retaining 'sequenced' supply arrangements by introducing a 'synchronised' phase, (based on the concept of demand or production transparency) for lower tiers. In the proposed 'synchronised' phase, production need not be 'in-sequence' but can be synchronised to the daily requirements of the vehicle assembly facility. Further upstream, batch production is retained for raw materials and components where variety is lower and production economies of scale prevail.

Three cases were undertaken to provide a thorough examination of the customer-driven concept. The cases provide real, test environments where the performance of separate value streams can be established and where an opportunity is created to assess the customer-driven concept through sharing vehicle production information across tiers based on the prototyping and simulation of an Internet-based, glass-pipeline arrangement (in each case), and through an audit of customer-driven processes (in the first case).

8.2 Case A

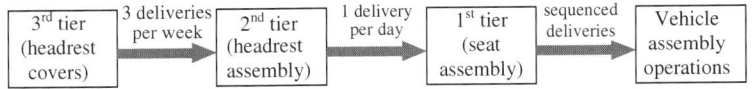

Fig. 8.2 Case A value stream structure

8.2 Case A

The vehicle assembler in case A produces entry-level luxury vehicles with annual production volumes in excess of 60,000 units. A team of researchers spent several months collecting and analysing relevant data at the case A assembly site, and exploring and developing the customer-driven approach. Figure 8.2 depicts the case value stream and the seat components chosen as the subject of the study. The flow of materials in case A follows a pattern characterised by three-weekly deliveries of components from the third to the second-tier supplier. The second-tier supplier delivers batches of 20 units on a daily basis. Seats are assembled just-in-sequence and delivered directly to the point-of-fit at the vehicle assembler. Options for vehicle seats include two-different styles (classic and sports), two types of material (cloth and leather), six different colours plus a choice of power, safety and adjustment features. The vehicle assembler and the first-tier seat assembler operate synchronously. The offset time between tiers upstream of the first tier is one day.

The total daily requirements for seat-sets are transmitted to the seat assembler via an EDI communications link. Each day the file provides the subsequent ten days' requirements, followed by a further forecast of demand in weekly and monthly time buckets. A work instruction for the assembly of a seat-set is initiated by the introduction of its associated vehicle into the final assembly sequence, at which point the actual seat requirement is communicated to the seat assembler via the EDI link. This communication is metronomic, triggered for every vehicle at a frequency equivalent to the final assembly takt time. The seat assembler uses the seat-set requirements' file as input to its own internal material requirements planning (MRP) system. The suppliers' schedules are produced for each of the seat assembler's component suppliers. Schedules consist of daily requirements for the following week, supplemented by weekly and monthly forecasts. The general flow of information for case A is depicted in Fig. 8.3.

The vehicle build plan is the result of actual dealer orders and the plant operating plan taking account of any rules concerning the prevention of sequencing violations. The vehicle build plan provides first-tier suppliers with daily requirements plus monthly forecasted production. The build plan was found to have poor stability and was subject to frequent changes.

In case A, the most popular seat option was referred to as the 'runner' and the least popular seat option was referred to as the 'stranger'. The average daily demand for the 'runner' was 81 units and the average daily demand for the 'stranger' was 4 units. The measurement of the performance of the value stream required the collection of data with respect to vehicle assembler (OEM) demand,

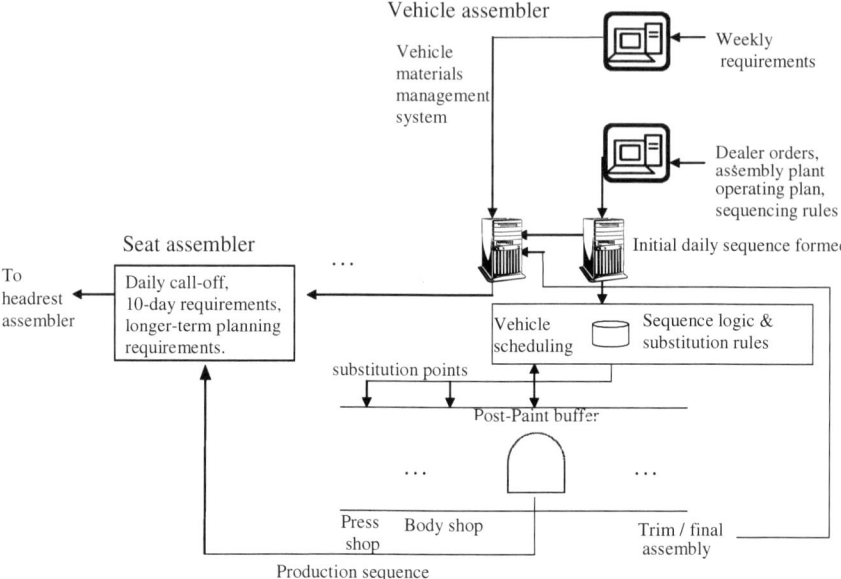

Fig. 8.3 Case A: Value stream information flow architecture

Table 8.1 Daily demand averages for case A: requirements data

Runner	D1	D2	D3	D4	D5	D6	D7	D8	D9	D10
OEM demand	106	84	86	98	43	104	91	89	128	0
Req to second tier	56	112	84	0	168	0	224	0	84	140
Deliveries third	0	200	0	140	0	80	200	0	320	0
Runner	D11	D12	D13	D14	D15	D16	D17	D18	D19	D20
OEM demand	131	98	88	75	0	90	92	66	58	91
Req to second tier	112	0	112	112	140	0	0	140	0	196
Deliveries third	0	240	0	240	0	0	80	0	0	0
Stranger	D1	D2	D3	D4	D5	D6	D7	D8	D9	D10
OEM demand	5	11	3	4	0	6	9	3	0	15
Req to second tier	0	0	0	0	0	0	0	0	28	0
Deliveries third	0	0	0	0	0	0	0	0	0	0
Stranger	D11	D12	D13	D14	D15	D16	D17	D18	D19	D20
OEM demand	8	7	1	2	0	3	5	1	1	0
Req to second tier	0	0	0	28	0	0	0	0	0	0
Deliveries third	0	0	20	0	0	0	0	0	0	0

requirements sent to the second-tier supplier and quantities delivered by the third tier. The data encompasses a period of 20 working days and is presented in Table 8.1.

8.2 Case A

Table 8.2 Case A value stream scorecard

	Runner	Stranger
Supply Chain Behaviour Measures		
Synchronisation index—overall (%)	38.6	32
First tier	96.0	96.0
Second tier	13.4	0
Third tier	6.4	0
Bullwhip measure	3.4	4.6
Responsiveness Measures		
Supply chain cycle times—overall (days)	9.2	17.4
First tier	2.3	10.5
Second tier	3.8	3.8
Third tier	3.1	3.1
Pipeline inventory—overall (days of inventory)	8.12	16.32
First tier components	1.82	10.02
Second tier finished goods	1.5	1.5
Second tier components	2.00	2.00
Third tier finished goods	2.8	2.8
Reliability Measures		
Stockout incidents—all value stream tiers	0	0
Backorders—all value stream tiers	0	0

A scorecard tool based upon the one presented in Chap. 7 was used to assess the performance of the 'runner' and 'stranger' options. The results, in terms of synchronisation index, bullwhip, supply chain cycle times and pipeline inventory, are presented in Table 8.2. The results revealed significant scope for improvements in all four metrics, particularly for the 'stranger' option. For example, low levels (less than 40%) of synchronisation were present, bullwhip deviated significantly from the optimum value of '1' and pipeline inventory could be improved by reducing component inventory levels at the first-tier supplier.

In order to examine customer-driven processes (the first principle of customer-driven supply chains) of the case A value stream, an audit based on the customer-driven guidelines displayed in Fig. 2.2 in Chap. 2 was undertaken. The content of the audit consisted of a series of statements that respondents were asked the extent to which they agreed with in relation to a particular value stream using a six-point Likert scale ranging from 'strongly disagree' (1) to 'strongly agree' (6). The audit statements have been transcribed in Table 8.3.

Figure 8.4 displays a series of ladder diagrams that represent the median responses of four production and engineering managers for the vehicle assembler for each of the four customer-driven guidelines (ordinate axis). (APD is the 'alignment of production with demand', IS is the 'integration of suppliers', EW is the 'elimination of waste' and CW is the 'creative involvement of the workforce'.) Each of the customer-driven guidelines has eight graduations, corresponding to the eight statements in the customer-driven audit (Table 8.3). Each ladder starts at the bottom with statement #1 and rises to the top with statement #32.

Table 8.3 Audit of customer-driven processes

Alignment of production with demand (APD)
1 Production is 'pulled' rather than 'pushed'.
2 Production is undertaken based on an instruction from a downstream process.
3 Production is regarded as 'make-to-downstream requirements' rather than 'make-to-stock'.
4 Production is paced to a customer demand rate or takt time (takt time is the customer demand rate and is calculated from the ratio of time available/day to demand/day).
5 Production rates vary in line with customer demand rates.
6 Production is mixed on the same processes and facilities.
7 Changes in demand volume and mix can be easily accommodated.
8 There is a commitment to reducing production run lengths and utilise a minimum economic batch size.

Integration of suppliers (IS)
9 Suppliers are actively supported in resolving their problems and improving performance.
10 Deliveries are based upon production requirements, are not excessive and arrive just before being used.
11 The notion of being a part of a complete (supply chain) value stream is both understood and accepted.
12 Suppliers receive schedules that are stable and predictable without unexpected changes.
13 Efforts are made to ensure that raw materials and ingredients are single sourced.
14 Suppliers have flexible processes that can easily accommodate demand changes.
15 Deliveries are made directly to the point-of-use (rather than to a remote storage area).
16 Supply inventory buffers are planned and set at minimum acceptable levels.

Elimination of waste (EW)
17 There is a real commitment to eliminate or minimise all non-value-adding activities.
18 Standard operating procedures are systematically used to provide work instructions.
19 Abnormal process behaviour is recognised and controlled by the workforce.
20 Visual displays are extensively used to support the standardisation and defect-free execution of the production process.
21 Quality systems and procedures are in place to prevent defects from moving to downstream operations.
22 The 5C system of workplace organisation is implemented and embraced by the workforce.
23 There is a commitment to reducing process set-up and changeover times.
24 TPM is well established and response to equipment breakdowns is systematised.

Creative involvement of the workforce (CW)
25 The workforce is actively involved in improvement activities and is empowered to make changes.
26 The work environment is organised so that most work is undertaken in teams.
27 Individual and team-based improvement ideas are regularly received from the workforce.
28 A structured programme of employee training is in place and adhered to.
29 Management devolve work-related decisions to production staff.
30 The workforce is multi-skilled and a system of job rotation is employed.
31 The work culture means that change is readily accepted and regarded as the norm.
32 Kaizen and constant, incremental improvement and innovation are embraced and practised by the workforce.

8.2 Case A

Fig. 8.4 Ladder diagrams for the vehicle assembler

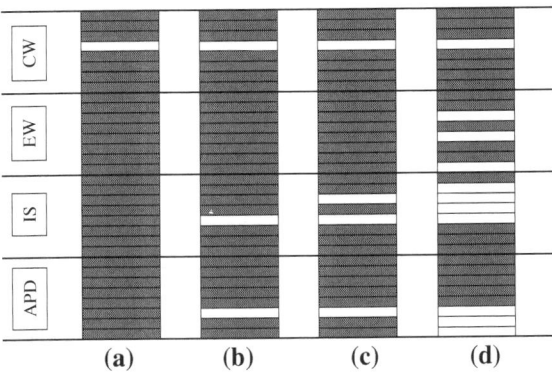

Fig. 8.5 Ladder diagrams for the seat assembler

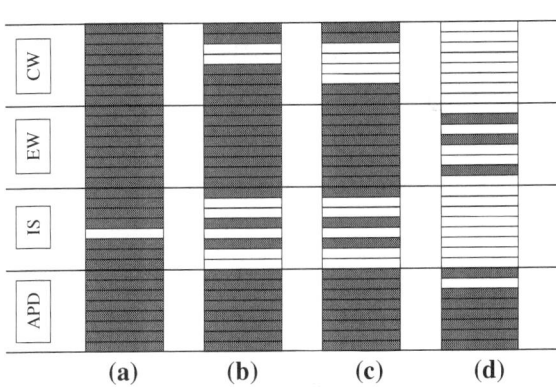

Graduations have been shaded in diagram (a) for responses with a median ≥4.00, in diagram (b) for responses with a median ≥4.50, in diagram (c) for responses with a median ≥5.00 and in diagram (d) for responses with a mean ≥5.50. In each case, the extent of the adoption of the customer-driven guidelines can be visually compared. Even at the 5.5 median level (Fig. 8.4d), a majority of customer-driven practices across each of the four guidelines is present. This reflects the maturity of the adoption of lean thinking practices generally in the high-volume automotive sector and specifically within this particular automotive plant. The creative involvement of the workforce is the guideline that is most populated by the respondents at the 5.5 level (Fig. 8.4d) despite 'management devolve work-related decisions to production staff' (audit statement #29) being the only creative involvement of the workforce practice that is not apparent at the median 4.0 level (Fig. 8.4a).

Figures 8.5, 8.6 and 8.7 display the corresponding ladder diagrams for the seat assembler, headrest assembler and the headrest material supplier (also referred to as the headrest cover or pocket supplier) respectively. In each case, four production and engineering staff were interviewed.

Figures 8.4, 8.5, 8.6 and 8.7 reveal inter-site comparisons of the extent to which customer-driven processes and practices have been embraced and adopted

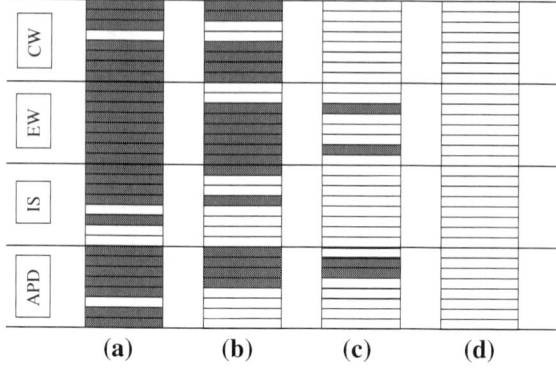

Fig. 8.6 Ladder diagrams for the headrest assembler

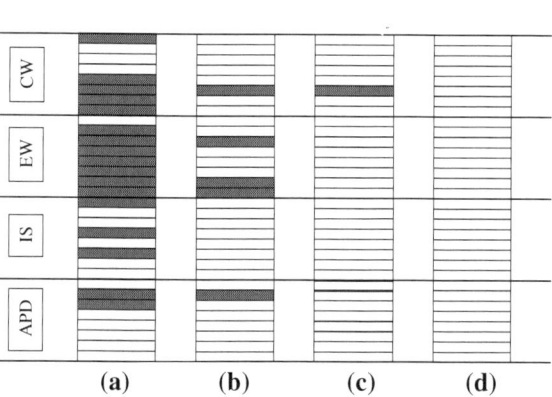

Fig. 8.7 Ladder diagrams for the headrest material supplier

throughout the value stream. Figure 8.5, the ladder diagram for the seat assembler, provides a general impression that customer-driven guidelines and practices are being adopted, but adoption is less than that for the vehicle assembler (Fig. 8.4) and is only evident at the highest 5.5 level for the 'alignment of production with demand (APD)'. This is as expected for a first-tier supplier which assembles in sequence. The 'white space', and therefore lack of adoption of customer-driven practices, is most conspicuous for the practices associated with the 'integration of suppliers (IS)'.

By contrast, the adoption of the customer-driven guidelines is far less apparent at the headrest assembler (Fig. 8.6) and headrest material supplier (Fig. 8.7). (It should be noted that these results are particular to this automotive value stream and even at the second and third-tier supply chain levels, the adoption of customer-driven practices may still be far higher than in other non-automotive, industry sectors.) The customer-driven guideline that most prominently features in Figs. 8.6 and 8.7 concerns the elimination of waste. All practices associated with this guideline achieved a median rating of 4 from the respondents for the headrest assembler and almost all practices associated with this guideline achieved a median rating of 4 for the headrest material supplier. Practices associated with the alignment of production with demand and the integration of suppliers are least

8.2 Case A

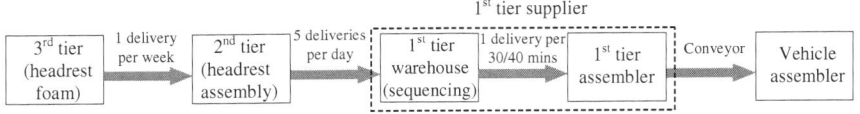

Fig. 8.8 Case B value stream structure

apparent at these upstream tiers. At the third tier, some of the supplier integration and alignment of production with demand practices, as portrayed in this study, are incompatible with the economies of scale necessary to drive efficient production and supply.

The four sets of ladder diagrams depict a downward trend in the adoption of customer-driven guidelines and practices from the vehicle assembler (OEM) to the headrest material supplier (third tier).

8.3 Case B

Case B concerns a high-volume (over 100,000) assembler of small, family vehicles. Seat options are less than in case A and consist of a choice of five different colours of cloth. Figure 8.8 depicts the structure of case B's value stream. Deliveries from the third-tier supplier are on a weekly basis. Deliveries from the second tier are made five times a day in batches of 56 units. Finished vehicle seats are transported via a conveyor direct to the point-of-fit at the vehicle assembly facility. The vehicle and seat assemblers produce synchronously according to a sequenced-supply arrangement. The offset time between tiers upstream of the first tier is one day.

In this case, aggregated daily seat requirements are transmitted to the seat assembler from the vehicle assembler. The seat requirements pertain to the next ten days, followed by tentative requirements for the subsequent weeks and months. In a similar manner to case A, a work instruction for the assembly of a seat-set is initiated by the introduction of its associated vehicle into the final assembly sequence, at which point the actual seat requirement is communicated to the seat assembler via the fibre optic EDI link. The seat assembler uses the seat requirements' file as an input to its internal MRP system. The file is 'loaded' each day, with the MRP system 'run' on a weekly basis. Schedules are produced for each of the second-tier component suppliers. These schedules contain daily requirements for the following week, and more tentative requirements for the subsequent weeks and months. The general flow of information for case B is the same as for case A.

Additionally in case B, the vehicle build plan, the result of the consolidation of direct dealer orders and the plant operating plan but also taking into account the facility's efficient mode of operation and its rules concerning the prevention of sequencing violations, provides the seat assembler's daily requirements for a

Table 8.4 Daily demand averages for case B: requirements data

Runner	D1	D2	D3	D4	D5	D6	D7	D8	D9	D10	D11
OEM demand	504	504	534	578	572	574	244	298	348	366	318
Req to Second tier	504	504	504	504	504	280	336	336	280	448	448
Deliveries third	2479	0	0	0	0	2479	0	0	0	0	2479
Runner	D12	D13	D14	D15	D16	D17	D18	D19	D20	D21	
OEM demand	416	510	612	700	418	470	780	522	534	554	
Req to second tier	448	448	448	504	560	504	560	560	560	504	
Deliveries third	0	0	0	0	2479	0	0	0	0	2479	
Stranger	D1	D2	D3	D4	D5	D6	D7	D8	D9	D10	D11
OEM demand	3	10	10	4	8	7	9	5	10	0	3
Req to second tier	0	0	0	0	112	0	0	0	0	0	0
Deliveries third	27	0	0	0	0	27	0	0	0	0	27
Stranger	D12	D13	D14	D15	D16	D17	D18	D19	D20	D21	
OEM demand	3	6	8	9	4	4	0	0	0	3	
Req to second tier	0	0	0	0	0	0	0	0	56	0	
Deliveries third	0	0	0	0	27	0	0	0	0	27	

ten-day horizon plus monthly forecasted production. The build plan was found to be prone to fluctuation until the final call-off.

The average daily demand for the most popular 'runner' and least popular 'stranger' options are 484 and 5 respectively. The measurement of the performance of the value stream required the collection of data for OEM demand, requirements sent to the second-tier supplier and quantities delivered by the third tier. The data covers a period of 21 working days for both the 'runner' and 'stranger' options (Table 8.4).

The scorecard tool was used to assess the performance of the 'runner' and 'stranger'options. The results, in terms of synchronisation index, bullwhip, supply chain cycle times and pipeline inventory, are shown in Table 8.5.

The results presented in Table 8.5 revealed that there is room for substantial improvements with respect to supply chain synchronisation, bullwhip effect, supply chain cycle times and pipeline inventory, particularly for the 'stranger' option. Synchronisation ranged from 33 to 60%, while bullwhip is high with values ranging between 3 and 7.5. Supply chain cycle times of 7.8 and 53 days were excessive.

8.4 Case C

In case C, the vehicle assembler offers three options for the seat-set of the vehicle studied. Figure 8.9 depicts case C's value stream and the modules/components chosen as the subject of the study. The flow of material between the second tier

8.4 Case C

Table 8.5 Case B value stream scorecard

	Runner	Stranger
Supply Chain Behaviour Measures		
Synchronisation index - overall (%)	59.63	33
First tier	99.0	99.0
Second tier	79.89	0
Third tier	0	0
Bullwhip measure	7.47	2.95
Responsiveness Measures		
Supply chain cycle times - overall (days)	7.79	53.0
First tier	2.08	47.36
Second tier	3.33	3.29
Third tier	2.38	2.35
Pipeline inventory—overall (days of inventory)	7.17	52.51
First tier components	2.04	47.33
Second tier finished goods	1.60	1.63
Second tier components	1.53	1.55
Third tier finished goods	2.00	2.00
Reliability Measures		
Stockout incidents—all value stream tiers	0	0
Backorders—all value stream tiers	0	0

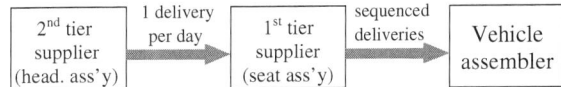

Fig. 8.9 Case C value stream structure

and the first tier comprises one daily delivery of a 72-unit batch. Similarly to cases A and B, the vehicle and seat assemblers operate synchronously using a sequenced arrangement. The offset time between tiers upstream of the first tier is one day.

Aggregated daily seat requirements are communicated to the seat supplier via a URL. This daily file provides the requirements (firm orders) for the next six days, followed by tentative requirements for the subsequent weeks and months. The seat-set assemby is initiated by the vehicle assembler fixing its final assembly sequence (the time window in case C is only three hours), at which time the actual seat requirement is communicated to the seat assembler from the post-Paint buffer at the vehicle assembly facility via an EDI broadcast system. The seat assembler uses the information to generate its materials' requirements. The spreadsheets containing the production schedules are sent to each of the component suppliers via e-mail. Generated schedules normally contain daily requirements for the following week plus tentative requirements for the subsequent weeks and months. The general flow of information for case C is depicted in Fig. 8.10. Vehicle build information provides the seat assembler's daily requirements plus monthly forecasted production. Quantities specified in the build plan are frozen from six days prior to the final call-off. The daily demand averages are 208 for the 'runner' option and 42 for the 'stranger' option.

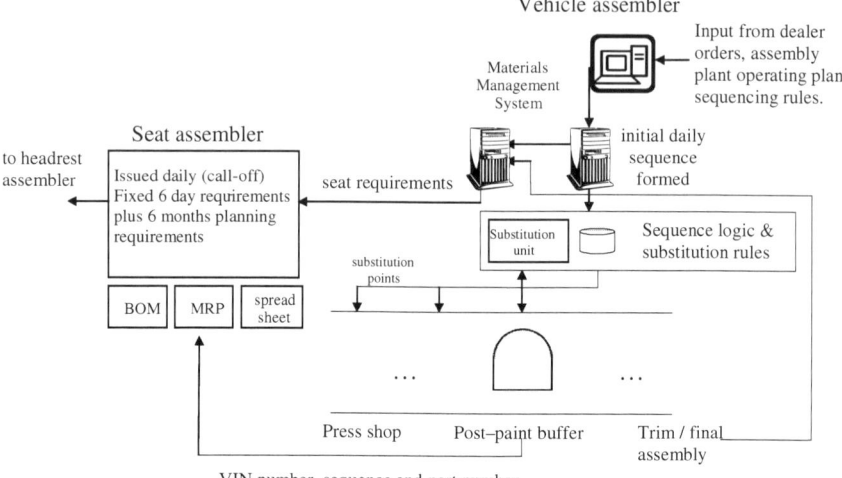

Fig. 8.10 Case C: Value stream information flow architecture

The assessment of the performance of the case C value stream required the collection of data for vehicle demand and quantities of headrests delivered by the second-tier supplier. The data presented in Tables 8.6 and 8.7 is for a period of 44 working days for both the 'runner' and 'stranger' options.

The results of the performance analysis are shown in Table 8.8. They reveal opportunities for improvements in supply chain behaviour and supply chain responsiveness measures. The table shows that both options carry similar pipeline inventories.

8.5 Improving Supply Chain Performance Through Glass Pipeline Design Changes

The value streams investigated in cases A, B and C follow a common convention in automotive supply chain design. That is, information generated by the OEM is visible to first but not second or third-tier suppliers. Therefore, it is not unexpected to find a variability in the demand signal amplified upstream in the chain. The performance of the value streams studied motivated the development of a prototype system based on a simple proposition: making OEM production requirements' information available to second and third-tier component suppliers would provide an opportunity for the performance of the value stream under consideration to be improved and the value streams in question to become aligned with the OEM demand signal. A secondary proposition was to regard an Internet-based arrangement as an appropriate means of making available the relevant information.

8.5 Improving Supply Chain Performance Through Glass Pipeline Design Changes

Table 8.6 Daily demand averages for case C: requirements data (runner)

Runner	D1	D2	D3	D4	D5	D6	D7	D8	D9	D10	D11
OEM demand	350	304	292	316	274	0	260	306	274	314	0
Deliv second tier	360	288	288	288	216	0	288	288	360	216	0
Runner	D12	D13	D14	D15	D16	D17	D18	D19	D20	D21	D22
OEM demand	212	190	194	210	246	0	200	200	196	196	244
Deliv to second tier	216	216	216	216	216	0	144	216	216	216	216
Runner	D23	D24	D25	D26	D27	D28	D29	D30	D31	D32	D33
OEM demand	0	210	186	208	228	252	0	260	192	200	192
Deliv second tier	0	144	216	216	216	216	0	216	216	144	216
Runner	D34	D35	D36	D37	D38	D39	D40	D41	D42	D43	D44
OEM demand	196	200	184	198	202	212	244	250	248	284	212
Deliv second tier	216	144	216	288	216	216	216	288	288	216	288

Table 8.7 Daily demand averages for case C: requirements data (stranger)

Stranger	D1	D2	D3	D4	D5	D6	D7	D8	D9	D10	D11
OEM demand	58	46	70	66	54	0	56	54	50	68	0
Deliv second tier	72	72	72	72	0	0	72	0	144	0	0
Stranger	D12	D13	D14	D15	D16	D17	D18	D19	D20	D21	D22
OEM demand	46	54	40	60	46	0	38	30	38	38	42
Deliv to second tier	72	72	72	72	0	0	0	72	0	72	0
Stranger	D23	D24	D25	D26	D27	D28	D29	D30	D31	D32	D33
OEM demand	0	42	36	36	34	48	0	38	56	36	54
Deliv second tier	0	72	0	72	0	72	0	0	72	72	0
Stranger	D34	D35	D36	D37	D38	D39	D40	D41	D42	D43	D44
OEM demand	34	50	48	54	46	52	46	54	42	50	36
Deliv second tier	72	72	72	0	72	72	72	0	72	72	0

Specifically, the prototype arrangements sought to establish the effect the upstream information-sharing schema devised has on:

- the capability of extending build-to-order (hence, customer-driven) production upstream in the chain,
- reducing the amplification between OEM demand fluctuations and those at the second and third tier,
- reducing the pipeline inventory held throughout the value stream,
- reducing the value stream cycle time.

Undertaking information sharing trials necessarily depends on using OEM-generated production requirements. Figure 8.11 depicts the outline structure of the

Table 8.8 Case C: scorecard results for the runner and stranger options

	Runner	Stranger
Supply Chain Behaviour Measures		
Synchronisation index - overall (%)	81.17	63.65
First tier	100	100
Second tier	62.35	27.30
Bullwhip measure	1.04	2.16
Responsiveness Measures		
Supply chain cycle times - overall (days)	4.98	5.80
First tier	1.39	2.20
Second tier	3.60	3.60
Pipeline inventory—overall (days of inventory)	4.85	5.67
First tier components	1.35	2.17
Second tier finished goods	1.50	1.50
Second tier components	2.00	2.00
Reliability Measures		
Stockout incidents—all value stream tiers	0	0
Backorders—all value stream tiers	0	0

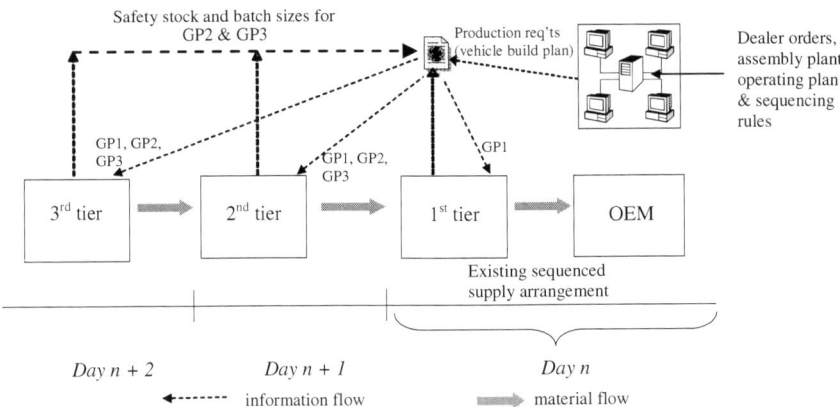

Fig. 8.11 Glass pipeline outline design (adapted from Coleman et al., 2004)

information sharing prototypes. These prototypes were designated the glass pipeline (GP) experiments and were conceived to provide enhanced visibility or transparency of information to upstream tiers.

The prototype design comprised three building blocks: an Information Stream, a Presentation Platform and a Viewer. Each building block consisted of several sub-sections that combined together to produce an output for the next block. A specialist Internet-service and solutions provider was used to provide web hosting for the necessary vehicle production requirements' data. In addition, an application was hosted on-line that could be configured to provide each business unit within a value stream with access to part requirements from the vehicle build plan. This

8.5 Improving Supply Chain Performance Through Glass Pipeline Design Changes 163

application was the central cortex of the new supply chain prototype designs. In order to provide vehicle build information in a format that the application could process, vehicle production data was transferred to a spreadsheet from which comma separated value (CSV) files were created and transmitted, using FTP, to the Internet provider's application. The application was hosted on a mainframe and each company involved in a case study was assigned an individual URL.

Three glass pipeline trials (GP1, GP2 and GP3) were undertaken. The 'glass pipeline basic mode' (GP1), provided data that was accessible to all tiers of the value stream. The resultant prototype architecture considered OEM production data and offset it by an appropriate lead time to create the second and third-tier demand requirements. The components chosen for each case were easily distinguished by colour and material type implying that second and third-tier requirements could be determined from the vehicle build plan, without the need to carry out a bill-of-materials (BOM) parts explosion.

The GP2 model used the vehicle build plan, on-hand pipeline inventory and delivery batch sizes as the required input to determine suitable second and third-tier requirements' information. A weekly 'look-up' reference to component inventory available at the first-tier supplier was used in the GP2 trial. The GP3 model was prototyped using the same approach as in GP2 but without the weekly reference to component inventory at the first-tier supplier.

In the GP1 trials, the requirements specified in the vehicle build plan were offset by one day, and made available to the second and third-tier suppliers. The implication of this is that the second-tier supplier would be required to deliver the production schedule requirements one full day before the vehicle assembly. The GP2 model used the same offset as GP1 to arrive at the raw demand for each tier on any given day. The GP2 algorithm compares available inventory with a pre-defined safety stock target level and reconciles this against raw demand to return available inventory to the desired level. The quantity used by the algorithm is rounded to the nearest pre-determined delivery batch size. Initial analysis retained both target inventory level and batch sizes at the same levels as the current system so that the magnitude of any improvement gained could be established. This arrangement demonstrated how inventory could be tightly controlled as second and third-tier synchronisation improved. Within this more tightly controlled regime, an analysis spreadsheet was used to simulate the system under different target inventory levels.

8.5.1 Glass Pipeline Trials for Case A

Figure 8.12 illustrates the prototype architecture devised for case A. The parameters used in the GP trials for case A are shown in Table 8.9.

Under this arrangement, all orders that had already been launched into build by the vehicle assembler were collated into a single file and presented to the suppliers (the suppliers access the file from the URL using a web browser) over a six-day

Fig. 8.12 Prototype architecture for case A

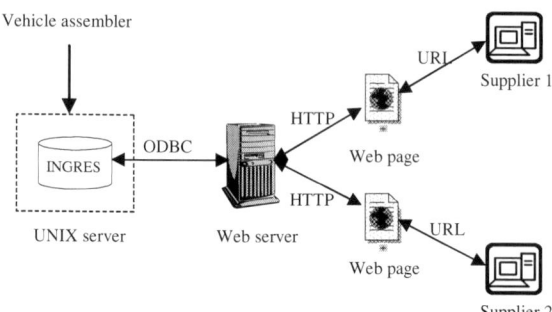

Table 8.9 GP trials' parameters for case A

Case A	Runner	Stranger
GP file data of origin	GP file	GP file
Resulting first tier component inventory GP1 (start)	102	14
Resulting second tier component inventory GP1 (start)	41	9
Resulting first tier component inventory GP2 & GP3 (start)	102	14
Resulting second tier component inventory GP2 & GP3 (start)	41	9
Desired first tier component inventory GP2 & GP3	74	12
Desired second tier component inventory GP2 & GP3	87	8
Batch sizes second tier GP2 & GP3	28	28
Batch sizes third tier GP2 & GP3	20	20

Table 8.10 Case A (R = runner and S = stranger) performance and component inventory scorecard

Case A	GP1 R	GP2 R	GP3 R	GP1 S	GP2 S	GP3 S
Behaviour Measures						
Synchronisation index—overall %	78.7	51.6	49.3	45.4	32.0	32.0
Bullwhip (OEM—second tier)	1.1	2.1	2.2	1.5	2.1	2.1
Responsiveness Measures						
Supply chain cycle times—overall (days)	5.7	6.6	6.2	8.8	12.0	11.5
Pipeline inventory—overall (days of inventory)	4.66	5.52	5.11	7.75	10.88	10.44
First tier component inventory (days)	1.87	1.06	0.98	3.80	3.77	4.43
Second tier component inventory (days)	0.21	1.01	1.12	1.37	3.67	3.0

production horizon. The system ran alongside the existing supply chain information and communication systems for a period of 20 weeks. Second and third-tier suppliers were asked to compare the quality and availability of the information received from the prototype system with the requirements' data obtained from the existing systems of communication. The results of the GP1, GP2 and GP3 trials for case A for both the the 'runner' and 'stranger' options are presented in Table 8.10.

8.5 Improving Supply Chain Performance Through Glass Pipeline Design Changes 165

Table 8.11 GP parameters for case B

Case B	Runner	Stranger
GP file data of origin	GP file	GP file
Resulting first tier component inventory GP1 (start)	1385	24
Resulting second tier component inventory GP1 (start)	470	11
Resulting first tier component inventory GP2 & GP3 (start)	190	10
Resulting second tier component inventory GP2 & GP3 (start)	190	4
Desired first tier component inventory GP2 & GP3	400	10
Desired second tier component inventory GP2 & GP3	430	10
Batch sizes second tier GP2 & GP3	56	7
Batch sizes third tier GP2 & GP3	10	10

Table 8.10 indicates that the GP trials yield improvements in component inventory held at the first and second tiers. The results for the 'runner' and 'stranger' options produced improvements to the current performance of the supply chain. Annex A presents the output from the GP trials for case A.

8.5.2 Glass Pipeline Trials for Case B

The GP trials for case B were undertaken using vehicle assembly schedules generated from three to six days prior to the final vehicle call-off. The GP prototype quantities were more closely aligned at the second tier with the final call-off than those specified on the output file of the first-tier MRP system. The value stream performance was compared with the results of the GP trials. The parameters used in the GP trials for case B are shown in Table 8.11.

Table 8.12 displays the results of the GP trials for case B. For the 'runner' option, GP1 provided the best performance in terms of synchronisation improvement and bullwhip reduction. The overall synchronisation index with GP1 improved from less than 40–90%. The bullwhip measure reduced from 7.47 to 1.00. The GP2 and GP3 trials significantly reduced the size of the starting inventory and the algorithm used for the GP2 and GP3 trials demonstrated reductions in pipeline inventory and supply chain cycle time. The results also indicate the improvements obtained for component inventory held at the first and second tier.

The GP trials for the 'stranger' option were simulated using vehicle requirements' data for one month. Table 8.12 demonstrates that pipeline inventory was reduced from 52.51 days of inventory to less than 7 days for each GP trial. Synchronisation was improved to at least 36% and bullwhip reduced to less than 1.60 on each occasion. The GP trials proved to be effective in reducing the excess inventory found at the first-tier supplier warehouse. The benefits of the GP trials become more evident when dealing with the 'stranger' option due to the very high inventory levels of components held by the first-tier supplier. Annex A presents the output from the GP trials for case B.

Table 8.12 Case B (R = runner and S = stranger) performance and component inventory scorecard

Case B	GP1 R	GP2 R	GP3 R	GP1 S	GP2 S	GP3 S
Behaviour Measures						
Synchronisation index—overall %	90.9	71.2	77.5	85.43	36.88	39.47
Bullwhip (OEM—second tier)	1.00	1.91	1.30	1.12	1.58	1.58
Responsiveness Measures						
Supply chain cycle times—overall (days)	5.16	4.81	4.53	6.02	6.98	6.71
Pipeline inventory—overall (days of inventory)	4.53	4.18	3.90	5.51	6.47	6.20
First tier component inventory (days)	1.50	0.60	0.68	2.45	1.68	1.75
Second tier component inventory (days)	0.86	0.70	0.70	0.88	1.88	1.91

Table 8.13 GP trials parameters for case C

Case C	Runner	Stranger
GP file data of origin	OEM req	OEM req
Resulting first tier component inventory GP1 (start)	351	59
Resulting second tier component inventory GP1 (start)	351	50
Resulting first tier component inventory GP2 & GP3 (start)	400	59
Resulting second tier component inventory GP2 & GP3 (start)	405	50
Desired first tier component inventory GP2 & GP3	280	50
Desired second tier component inventory GP2 & GP3	280	86
Batch sizes second tier GP2 & GP3	36	7

8.5.3 Glass Pipeline Trials for Case C

The glass pipeline trials for case C were undertaken using a six-day vehicle assembly schedule. The performance of the case C value stream was compared with the outcome of the glass pipeline trials using the same scorecard tool that was used to articulate performance in cases A and B. The parameters used in the GP trials for case C are shown in Table 8.13. Each of the GP trials was simulated using production quantities for eight weeks. Table 8.14 displays the results. Trials with the 'runner' option revealed the scope to increase synchronisation (100% for GP1, 83.91% for GP2 and 82.29% for GP3), reduce bullwhip effect (less than 1.74 for each GP trial) and reduce pipeline inventory and supply chain cycle times.

The results for the 'stranger' option include synchronisation improvement (100% for GP1, 75% for GP2 and GP3), bullwhip reduction (less than 2 for all GP trials), and reduction in pipeline inventory and supply chain cycle time. Furthermore, Table 8.14 displays the extent of the improvements expected in case C for component inventory held at the first and second-tier suppliers. Annex A presents the output from the GP trials for case C.

8.5 Improving Supply Chain Performance Through Glass Pipeline Design Changes 167

Table 8.14 Case C (R = runner and S = stranger) performance and component inventory scorecard

Case C	GP1 R	GP2 R	GP3 R	GP1 S	GP2 S	GP3 S
Behaviour Measures						
Synchronisation index—overall %	100	83.91	82.29	100	75.57	75.35
Bullwhip (OEM—second tier)	1	1.61	1.74	1	1.82	1.88
Responsiveness Measures						
Supply chain cycle times—overall (days)	3.21	4.16	4.06	2.89	4.53	4.60
Pipeline inventory—overall (days of inventory)	3.08	4.03	3.93	2.75	4.40	4.47
First tier component inventory (days)	0.97	1.41	1.34	0.99	1.21	1.23
Second tier component inventory (days)	1.15	1.34	1.34	1.04	2.02	2.05

8.5.4 Discussion of Glass Pipeline Trials

In each case, the scorecard demonstrated how information sharing and glass pipeline designs can improve value stream performance. In the GP1 trials, batches were replaced with one-to-one deliveries and produced the most dramatic improvements but without recognition of the costs involved in implementing new work agreements between tiers of suppliers and the OEMs. The GP2 and GP3 trials used a conventional batch arrangement but component inventory levels were only taken into account with GP2. The results demonstrated potential improvements but, unlike GP1, were not seriously affected by the reliability of the data source available. Case C was the best-performing supply arrangement although its results need to be qualified by the omission of the third-tier analysis due to unavailability of appropriate data.

8.6 Analysis of Customer-Driven Processes

The customer-driven process profile for the vehicle assembler in case A was not expected to change significantly as a result of the glass pipeline trials as most customer-driven practices had already been adopted. Discussions with the original respondents of the customer-driven process questionnaire (Table 8.3) suggested that only the 'integration of suppliers (IS)' customer-driven guideline would expect some perceptible change as a direct result of the glass pipeline trials. Figure 8.13 displays a corresponding series of ladder diagrams to those in Fig. 8.4. Again, the ladders represent the median responses of the four production and engineering managers for the vehicle assembler for each of the four customer-driven guidelines and graduations have been shaded in diagram (a) for responses with a median ≥ 4.00, in diagram (b) for responses with a median ≥ 4.50, in diagram (c) for responses with a median ≥ 5.00 and in diagram (d) for responses with a mean ≥ 5.50. On this occasion, however, the ladders provide a 'to be'

Fig. 8.13 Expected ladder diagrams for the vehicle assembler after GP trials

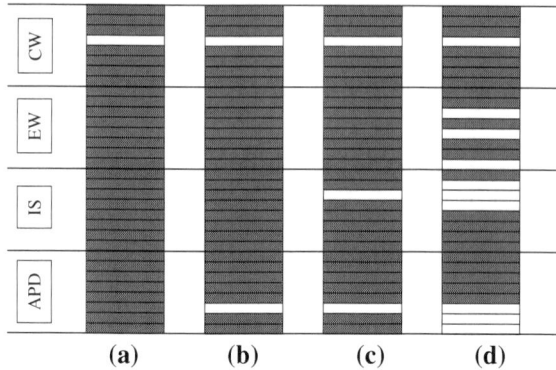

representation based upon a consolidated view of the three glass pipeline trials. In Fig. 8.13, some improvement can be seen in the 'IS' profile particularly in relation to practice point 12—'suppliers receive schedules that are stable and predictable without unexpected changes'. A glass-pipeline, information-sharing implementation would provide the requisite motivation to improve schedule stability.

Discussions with the management and production team at the seat assembler suggested the following:

- The 'alignment of production with demand (APD)', 'elimination of waste (EW)' and 'creative involvement of the workforce (CW)' guidelines would not be expected to change as a result of the glass pipeline trials. However, several of the practices associated with the 'integration of suppliers (IS)' would improve as a result of each of the trials.
- Each trial would significantly improve the median position of practice point 12—'suppliers receive schedules that are stable and predictable without unexpected changes'. The disruptive effects of regular schedule changes would be eliminated with the glass pipeline information sharing design.
- Point 10—'deliveries are based upon production requirements, are not excessive and arrive just before being used' was also seen as improving with the most conspicuous improvement being associated with GP1.
- Point 14—'suppliers have flexible processes that can easily accommodate demand changes' was also seen as improving as a result of the trials.

Figure 8.14 displays a corresponding series of ladder diagrams to those in Fig. 8.5. Clear improvement can be seen in the 'IS' profile.

Discussions with the management and production team at the headrest assembler suggested the following:

- The 'creative involvement of the workforce (CW)' guideline would not be expected to change as a direct result of the glass pipeline trials for the headrest assembler. Only practice point 23—'there is a commitment to reducing process set-up and changeover times', from the 'elimination of waste (EW)' guideline,

8.6 Analysis of Customer-Driven Processes

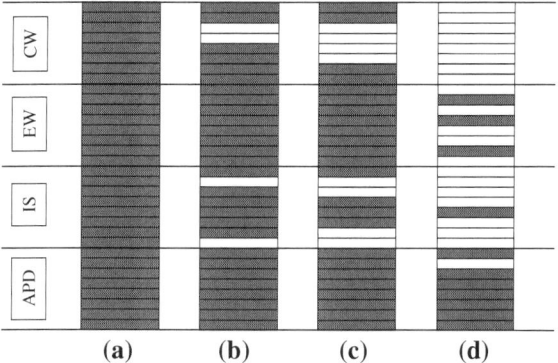

Fig. 8.14 Expected ladder diagrams for the seat assembler after GP trials

showed some improvement motivated by the enhanced information integration of the glass pipeline design. However, several of the practices associated with the 'alignment of production with demand (APD)' and the 'integration of suppliers (IS)' would improve as a result of each of the trials with a general reduction in the 'white space' for both the IS and APD practice profiles.

- Each trial would improve the median position of six of the eight APD practice points. Production would be more customer-order and takt-driven than the prevailing 'as is' situation.
- Similarly to the seat assembler, each trial would significantly improve the median position of practice point 12—'suppliers receive schedules that are stable and predictable without unexpected changes'. This obviates the need for large buffer inventories and explains the rationale for the subsequent improvement in the APD profile.
- The respondents' perception that the tighter information integration would bring about a significant improvement in practice point 11—'the notion of being a part of a complete (supply chain) value stream is both understood and accepted'.

Figure 8.15 displays a corresponding series of ladder diagrams for the headrest assembler to those in Fig. 8.6. Discussions with the management and production team at the headrest material supplier suggested the there would be a general improvement in the APD and IS profiles. Figure 8.16 displays a corresponding series of ladder diagrams for the headrest material supplier to those in Fig. 8.7.

8.7 Extending Synchronisation Upstream in the Supply Chain

A further investigation concerned a study of the feasibility of extending the sequence boundary to the second tier of suppliers. This requires the communication of the sequence broadcast to a headrest supplier in case studies A, B and C.

Fig. 8.15 Expected ladder diagrams for the headrest assembler after GP trials

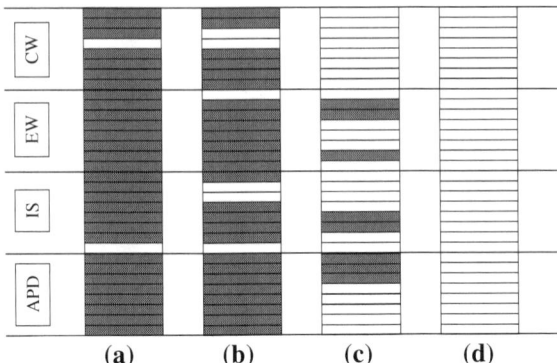

Fig. 8.16 Expected ladder diagrams for the headrest material supplier after GP trials

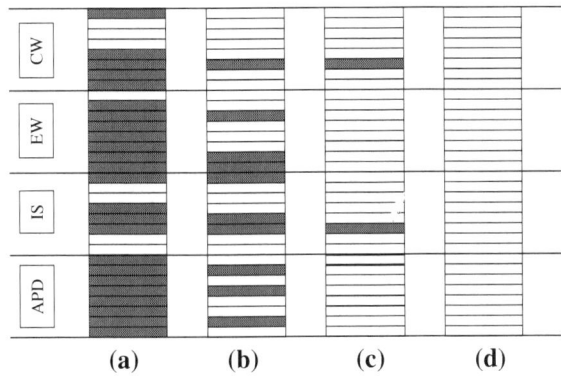

Figure 8.17 depicts a simple illustration of the information and material flow architecture for the proposed arrangement.

Figures 8.18 and 8.19 depict how synchronisation levels in the supply chain can be increased by extending the sequence boundary to a second-tier supplier. Such an arrangement extends the 'metronomic', build-to-sequence portion of the value stream, provides an opportunity for a deeper level of product customisation and improves the customer-driven design and behaviour of the supply chain. The opportunity to consider the feasibility of such an arrangement is dependent on the geographic proximity between the OEM and the first and second-tier suppliers.

8.7.1 Approach to Extending Synchronisation Beyond a First Tier

Case C was chosen as the subject of the study as both the first and second-tier suppliers were located in close proximity to the vehicle assembler. The parameters considered in the proposed solution are displayed in Table 8.15. Vehicle assembly

8.7 Extending Synchronisation Upstream in the Supply Chain

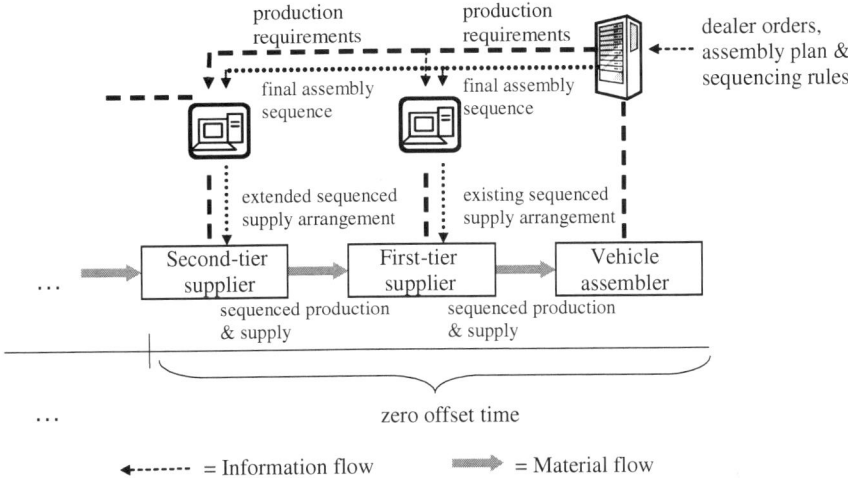

Fig. 8.17 Extending the sequence boundary to the second tier

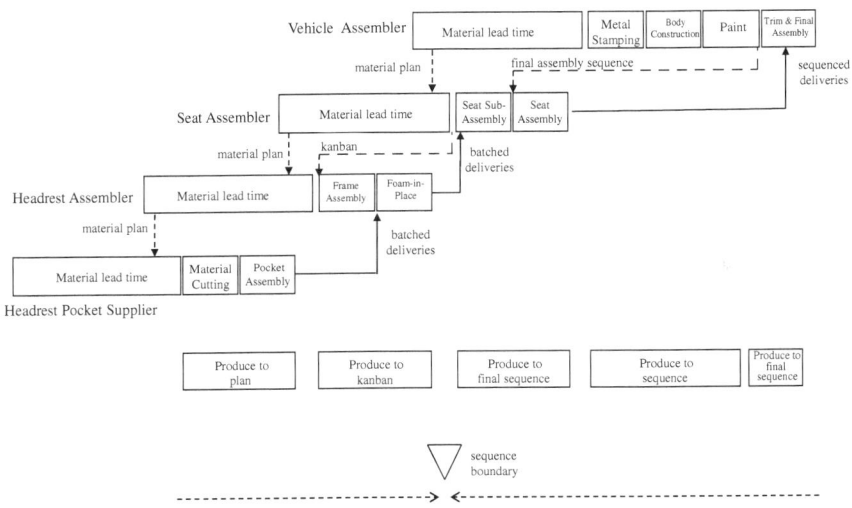

Fig. 8.18 General representation of the sequenced supply arrangement

operations' demand, on-hand pipeline inventory and delivery batch sizes were used to determine second and third-tier requirements. All references are made to inventory levels adopted by the proposed solution.

The proposed approach uses the demand requirements specified in the vehicle assembler's production files to determine the raw demand for each tier on a given day. The algorithm compares available inventory (C_0 and D_0) with a pre-defined safety stock target level (*aa* and *ab*) and adds or subtracts from the raw demand to

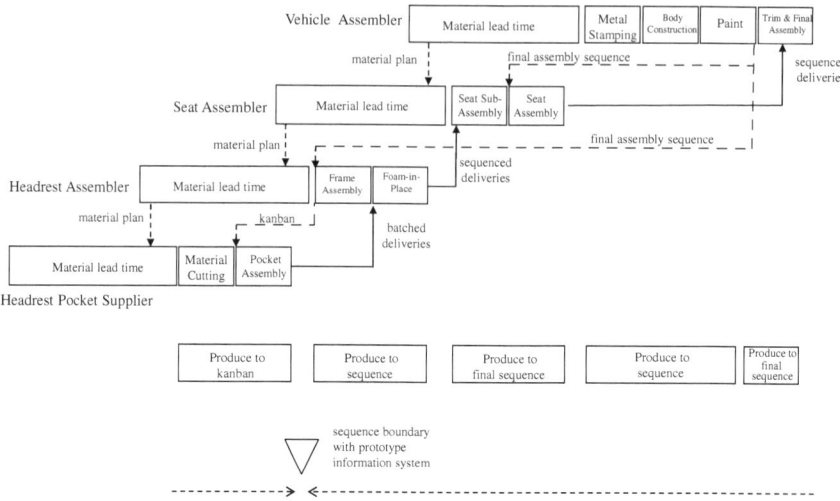

Fig. 8.19 General representation of the second-tier sequenced supply arrangement

Table 8.15 Elements of the second-tier sequencing solution

n	Period examined (days)
R_n	Already available fixed daily OEM raw demand
aa	Predefined first-tier component inventory level
ab	Predefined second-tier component inventory level
ba	Second-tier batch size deliveries (units)
bb	Third-tier batch size deliveries (units)
C_0	Initial first-tier component inventory
D_0	Initial second-tier component inventory
S_n	Second-tier component deliveries
F_n	Third-tier component deliveries

return available inventory to the desired level. The quantity used by the algorithm is rounded to the nearest pre-determined delivery batch size. Initial analysis maintained both the target inventory level and the batch sizes the same as the current system in order to better appreciate the effect of implementing a system of second-tier sequencing. The pseudo-code of the algorithm used is presented in Fig. 8.20 (also, see Mondragon and Lyons 2008).

8.7.2 The Impact of Implementing Second-Tier Sequencing in Case C

An implication of the proposed change concerned the elimination of the offset time of one day between the second and first tiers and changing second-tier batch size

8.7 Extending Synchronisation Upstream in the Supply Chain

```
i = 1
l = 1
**/ second-tier material deliveries /**
For i =1 to n step 1
{
                if aa + Rn – Cn-1 < 0
                  Sn = 0
                else
                  Sn = round up (aa + Rn – Cn-1, ba)
                Cn = Cn-1 + Sn – Rn
}
**/ third-tier material deliveries /**
For I = 1 to n step 1
{
                if ab + Rn+1 – Dn-1< 0
                  Tn = 0
                else
                  Tn = round up(ab+ Rn+1 – Dn-1, bb) **/one day offset time between 3rd and 2nd tiers/**
                Dn = Dn-1 + Tn – Rn+1
                n = n+1
}
```

Fig. 8.20 Algorithm employed for second-tier sequenced deliveries

deliveries to one. Changes to build-to-order capability and make-to-sequence are depicted in Fig. 8.21. The shaded portions of the figure indicate those processes that are driven by the final sequence of the vehicle assembler. In the right-hand diagram the shaded processes extend to the second-tier headrest assembly processes.

The stranger trim option was used in the trials (42 units per day). A summary of the results of modelling the implementation of second-tier sequenced deliveries are shown in Table 8.16. The output of the algorithm for case C is shown in Annex B. The values shown in Table 8.16 indicate that with the introduction of a sequenced second tier it is possible to record a perfect bullwhip index of 1.00—the variance of the demand recorded at the second-tier (shown as deliveries) is equal to that recorded at the point of origin.

8.8 Chapter Summary

The chapter has attempted to demonstrate how the principles of customer-driven supply chain design in the form of synchronised planning, production and delivery activities between supply chain tiers, and the assessment and improvement of customer-driven processes can be achieved through the provision of a glass pipeline system for information sharing. The glass pipeline concept makes demand-related information accessible upstream in a supply chain. It helps to dampen the amplification which occurs when information is transmitted

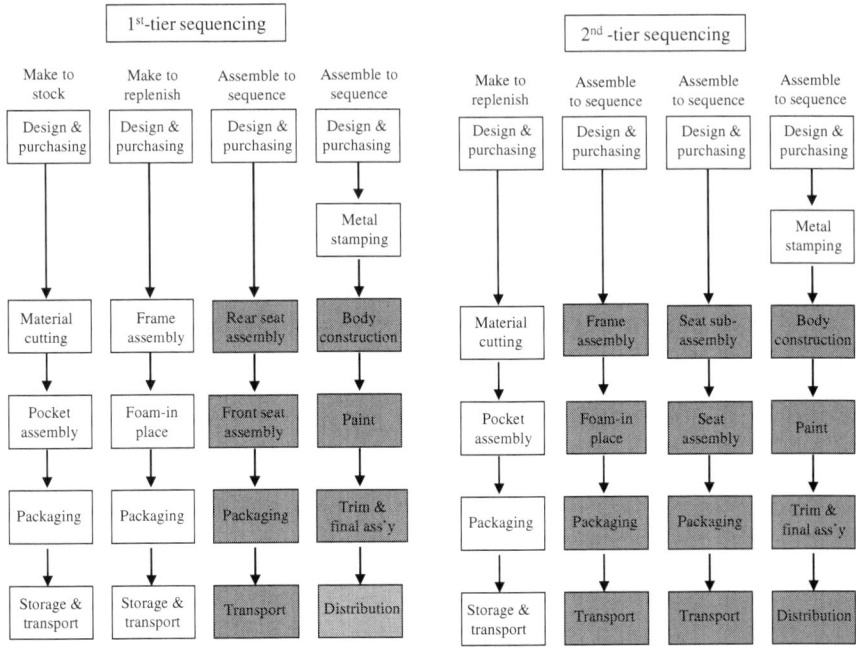

Fig. 8.21 The impact of sequencing on build-to-order capability

Table 8.16 Performance analysis for Case C

	Second tier	First tier	Overall at OEM level
Deliveries downstream	Sequenced	Sequenced	–
Synchronisation index	100%	100%	100%
Bullwhip index	–	–	1.00
Pipeline Inventory (days)	Components: 1.86 days Finished goods: 0 days	Components: 0.02 days Finished goods: 0 days	1.88 days
Stockouts/Backorders	0/0	0/0	–

sequentially to upstream tiers and provides suppliers with an opportunity to more closely align production and delivery with customer demand.

Each of the three automotive cases demonstrated how the glass pipeline concept allied to the inculcation of a range of customer-driven guidelines and practices can be used to facilitate a customer-driven supply chain that is capable of managing and co-ordinating low-quantity, high-variety requirements and, when necessary, provide customised items in single units at a high degree of efficiency. For each case, the prototype information architecture made OEM production requirements accessible via a specialist third-party provider for suppliers to view on the Internet. The content of the site was created using open Internet standards but required

8.8 Chapter Summary

configuring and using a web browser and kept up-to-date on a daily basis with transmitted information via FTP. The availability of a forward, visible workload resulted in more predictable production schedules, reduced safety stocks, and improved responsiveness, co-ordination and synchronisation.

Additionally, in case A, the glass pipeline trials had a significant, positive impact on the customer-driven practices particularly those relating to the 'alignment of production with demand' and the 'integration of suppliers'. Furthermore, in case C, the results of extending synchronisation upstream to a second-tier supplier showed that the same benefits of synchronisation experienced between an OEM and a first-tier supplier are mimicked upstream in the chain. A supply chain that includes second-tier synchronised sequencing demands close proximity between tiers in addition to an appropriate glass pipeline but can substantially enhance supply chain performance in terms of bullwhip reduction, overall synchronisation levels and reduced supply chain inventories and cycle times whilst avoiding backorders and stockout incidents.

Annex A: Output from the GP Trials

Case A—medium volume supply chain. Runner

GP composite file requirements list

Day	D-1	D-2	D-3	D-4	D-44	D-6	D-7
Requirements	200	115	83	104	200	118	95
Day	D-8	D-9	D-10	D-11	D-12	D-13	D-14
Requirements	95	95	122	122	114	103	100
Day	D-15	D-16	D-17	D-18	D-19	D-20	D-21
Requirements	0	80	71	83	73	0	63

Final call-off file requirements list

Day	D-1	D-2	D-3	D-4	D-44	D-6	D-7
Requirements	106	84	86	98	43	104	91
Day	D-8	D-9	D-10	D-11	D-12	D-13	D-14
Requirements	89	128	0	131	98	88	75
Day	D-15	D-16	D-17	D-18	D-19	D-20	D-21
Requirements	0	90	92	66	58	91	–

GP1 Results, Case A, Runner

Day	D-1	D-2	D-3	D-4	D-44	D-6	D-7
Deliveries second tier	115	83	104	1	118	95	95
Deliveries third tier	83	104	1	118	95	95	122
First tier component inventory	111	110	128	31	106	97	101
Second tier component inventory	9	30	-73	44	21	21	48
Day	D-8	D-9	D-10	D-11	D-12	D-13	D-14
Deliveries second tier	122	122	114	103	100	85	0
Deliveries third tier	122	114	103	100	85	0	80
First tier component inventory	134	128	242	214	216	213	138
Seond tier component inventory	48	40	29	26	11	-74	6

GP1 Results, Case A, Runner

Day	D-15	D-16	D-17	D-18	D-19	D-20	D-21
Deliveries second tier	80	71	83	73	0	63	–
Deliveries third tier	71	83	73	0	63	–	–
First tier component inventory	218	199	190	197	139	111	–
Second tier component inventory	-3	9	-1	-74	-11	-74	–

GP2 Results, Case A, Runner

Day	D-1	D-2	D-3	D-4	D-44	D-6	D-7
Deliveries second tier	84	84	112	0	196	112	0
Deliveries third tier	120	120	0	200	0	100	220
First tier component inventory	80	80	106	8	161	169	78
Second tier component inventory	77	113	1	201	5	-7	213
Day	D-8	D-9	D-10	D-11	D-12	D-13	D-14
Deliveries second tier	112	84	140	0	112	84	0
Deliveries third tier	0	100	80	100	20	20	160
First tier component inventory	101	57	197	66	80	76	1
Second tier component inventory	101	117	57	157	65	1	161
Day	D-15	D-16	D-17	D-18	D-19	D-20	D-21
Deliveries second tier	140	28	84	84	0	112	–
Deliveries third tier	0	100	60	20	140	0	–
First tier component inventory	141	79	71	89	31	52	–
Second tier component inventory	21	93	69	5	145	33	–

Annex A: Output from the GP Trials

GP3 Results, Case A, Runner

Day	D-1	D-2	D-3	D-4	D-44	D-6	D-7
Deliveries second tier	84	84	112	0	196	112	0
Deliveries third tier	120	120	0	200	0	100	220
First tier component inventory	80	80	106	8	161	169	78
Second tier component inventory	77	113	1	201	5	-7	213

GP3 Results, Case A, Runner, continued

Day	D-8	D-9	D-10	D-11	D-12	D-13	D-14
Deliveries second tier	112	84	140	0	112	84	0
Deliveries third tier	120	100	80	140	0	0	160
First tier component inventory	101	57	197	66	80	76	1
Second tier component inventory	101	117	57	197	85	1	161
Day	D-15	D-16	D-17	D-18	D-19	D-20	D-21
Deliveries second tier	140	0	112	84	0	112	–
Deliveries third tier	0	140	0	40	140	0	–
First tier component inventory	141	51	71	89	31	52	–
Second tier component inventory	21	161	49	5	145	33	–

CASE A—Medium volume supply chain. Stranger

GP composite file requirements list

Day	D-1	D-2	D-3	D-4	D-44	D-6	D-7
Requirements	3	5	13	1	0	6	3
Day	D-8	D-9	D-10	D-11	D-12	D-13	D-14
Requirements	3	0	23	0	17	4	3
Day	D-15	D-16	D-17	D-18	D-19	D-20	D-21
Requirements	0	1	2	6	0	0	1

Final call-off file requirements list

Day	D-1	D-2	D-3	D-4	D-44	D-6	D-7
Requirements	5	11	3	4	0	6	9
Day	D-8	D-9	D-10	D-11	D-12	D-13	D-14
Requirements	3	0	15	8	7	1	2
Day	D-15	D-16	D-17	D-18	D-19	D-20	D-21
Requirements	0	3	5	1	1	0	–

GP1 Results, Case A, Stranger

Day	D-1	D-2	D-3	D-4	D-44	D-6	D-7
Deliveries second tier	5	13	1	0	6	3	3
Deliveries third tier	13	1	0	6	0	3	0
First tier component inventory	14	16	14	10	16	13	7
Second tier component inventory	17	5	4	10	4	4	1
Day	D-8	D-9	D-10	D-11	D-12	D-13	D-14
Deliveries second tier	0	23	0	17	4	3	0
Deliveries third tier	23	0	17	4	3	0	1
First tier component inventory	4	27	12	21	18	20	18
Second tier component inventory	24	1	18	5	4	1	2
Day	D-15	D-16	D-17	D-18	D-19	D-20	D-21
Deliveries second tier	1	2	6	0	0	1	–
Deliveries third tier	2	6	0	0	1	–	–
First tier component inventory	19	18	19	18	17	18	–
Second tier component inventory	3	7	1	1	2	1	–

GP2 Results, Case A, Stranger

Day	D-1	D-2	D-3	D-4	D-44	D-6	D-7
Deliveries second tier	0	28	0	0	0	0	0
Deliveries third tier	20	0	0	20	0	0	0
First tier component inventory	9	26	23	19	19	13	4
Second tier component inventory	29	1	1	21	21	21	21
Day	D-8	D-9	D-10	D-11	D-12	D-13	D-14
Deliveries second tier	0	28	0	0	0	28	0
Deliveries third tier	20	0	20	0	0	0	0
First tier component inventory	1	29	14	6	-1	26	24
Second tier component inventory	41	13	33	33	33	5	5

GP2 Results, Case A, Stranger, continued

Day	D-15	D-16	D-17	D-18	D-19	D-20	D-21
Deliveries second tier	0	0	0	0	0	0	–
Deliveries third tier	0	0	0	0	0	0	–
First tier component inventory	24	21	16	15	14	14	–
Second tier component inventory	5	5	5	5	5	5	–

Annex A: Output from the GP Trials

GP3 Results, Case A, Stranger

Day	D-1	D-2	D-3	D-4	D-44	D-6	D-7
Deliveries second tier	0	28	0	0	0	0	0
Deliveries third tier	20	0	0	20	0	0	0
First tier component inventory	9	26	23	19	19	13	4
Second tier component inventory	29	1	1	21	21	21	21
Day	D-8	D-9	D-10	D-11	D-12	D-13	D-14
Deliveries second tier	0	28	0	28	0	0	0
Deliveries third tier	20	0	20	0	0	0	0
First tier component inventory	1	29	14	34	27	26	24
Second tier component inventory	41	13	33	5	5	5	5
Day	D-15	D-16	D-17	D-18	D-19	D-20	D-21
Deliveries second tier	0	0	0	0	0	0	–
Deliveries third tier	0	0	0	0	0	0	–
First tier component inventory	24	21	16	15	14	14	–
Second tier component inventory	5	5	5	5	5	5	–

CASE B – High volume supply chain. Runner

GP composite file requirements list

Day	D-1	D-2	D-3	D-4	D-44	D-6	D-7
Requirements	454	452	510	488	512	240	298
Day	D-8	D-9	D-10	D-11	D-12	D-13	D-14
Requirements	334	340	288	416	416	412	418
Day	D-15	D-16	D-17	D-18	D-19	D-20	D-21
Requirements	418	470	528	522	528	560	318

Final call-off file requirements list

Day	D-1	D-2	D-3	D-4	D-44	D-6	D-7
Requirements	504	534	578	572	574	244	298
Day	D-8	D-9	D-10	D-11	D-12	D-13	D-14
Requirements	348	366	318	416	510	612	700
Day	D-15	D-16	D-17	D-18	D-19	D-20	D-21
Requirements	418	470	780	522	534	554	318

GP1 Results, Case B, Runner

Day	D-1	D-2	D-3	D-4	D-44	D-6	D-7
Deliveries second tier	452	510	488	512	240	298	334
Deliveries third tier	510	488	512	240	298	334	340
First tier component inventory	1333	1309	1219	1159	825	879	915
Second tier component inventory	528	506	530	258	316	352	358
Day	D-8	D-9	D-10	D-11	D-12	D-13	D-14
Deliveries second tier	340	288	416	416	412	418	418
Deliveries third tier	288	416	416	412	418	418	470
First tier component inventory	907	829	927	927	829	635	353
Second tier component inventory	306	434	434	430	436	436	488

GP1 Results, Case B, Runner

Day	D-15	D-16	D-17	D-18	D-19	D-20	D-21
Deliveries second tier	470	528	522	528	560	318	–
Deliveries third tier	528	522	528	560	318		–
First tier component inventory	405	463	205	211	237	1	–
Second tier component inventory	546	540	546	578	336	18	–

GP2 Results, Case B, Runner

Day	D-1	D-2	D-3	D-4	D-44	D-6	D-7
Deliveries second tier	672	560	504	616	280	616	280
Deliveries third tier	750	650	580	240	670	320	620
First tier component inventory	358	384	310	354	60	432	414
Second tier component inventory	268	358	434	58	448	152	492
Day	D-8	D-9	D-10	D-11	D-12	D-13	D-14
Deliveries second tier	336	280	504	0	728	504	616
Deliveries third tier	230	460	280	500	10	720	560
First tier component inventory	402	316	502	86	304	196	112
Second tier component inventory	386	566	342	842	124	340	284
Day	D-15	D-16	D-17	D-18	D-19	D-20	D-21
Deliveries second tier	784	0	896	784	560	280	–
Deliveries third tier	670	780	10	930	540	240	–
First tier component inventory	478	8	124	386	412	138	–
Second tier component inventory	170	950	64	210	190	150	–

Annex A: Output from the GP Trials

GP3 Results, Case B, Runner

Day	D-1	D-2	D-3	D-4	D-44	D-6	D-7
Deliveries second tier	672	560	504	616	280	616	280
Deliveries third tier	750	650	580	240	670	320	620
First tier component inventory	358	384	310	354	60	432	414
Second tier component inventory	268	358	434	58	448	152	492

GP3 Results, Case B, Runner, continued

Day	D-8	D-9	D-10	D-11	D-12	D-13	D-14
Deliveries second tier	336	280	504	336	392	504	616
Deliveries third tier	230	460	280	500	340	390	560
First tier component inventory	402	316	502	422	304	196	112
Second tier component inventory	386	566	342	506	454	340	284
Day	D-15	D-16	D-17	D-18	D-19	D-20	D-21
Deliveries second tier	784	448	448	784	560	280	–
Deliveries third tier	670	780	460	480	540	240	–
First tier component inventory	478	456	124	386	412	138	–
Second tier component inventory	170	502	514	210	190	150	–

Case B—High volume supply chain. Stranger

GP composite file requirements list

Day	D-1	D-2	D-3	D-4	D-44	D-6	D-7
Requirements	3	10	6	2	7	5	8
Day	D-8	D-9	D-10	D-11	D-12	D-13	D-14
Requirements	4	8	0	3	3	4	5
Day	D-15	D-16	D-17	D-18	D-19	D-20	D-21
Requirements	9	4	1	1	0	0	1

Final call-off file requirements list

Day	D-1	D-2	D-3	D-4	D-44	D-6	D-7
Requirements	3	10	10	4	8	7	9
Day	D-8	D-9	D-10	D-11	D-12	D-13	D-14
Requirements	5	10	0	3	3	6	8
Day	D-15	D-16	D-17	D-18	D-19	D-20	D-21
Requirements	9	4	4	0	0	0	1

GP1 Results, Case B, Stranger

Day	D-1	D-2	D-3	D-4	D-44	D-6	D-7
Deliveries second tier	10	6	2	7	5	8	4
Deliveries third tier	6	2	7	5	8	4	8
First tier component inventory	31	27	19	22	19	20	15
Second tier component inventory	7	3	8	6	9	5	9
Day	D-8	D-9	D-10	D-11	D-12	D-13	D-14
Deliveries second tier	8	0	3	3	4	5	9
Deliveries third tier	0	3	3	4	5	9	4
First tier component inventory	18	8	11	11	12	11	12
Second tier component inventory	1	4	4	5	6	10	5
Day	D-15	D-16	D-17	D-18	D-19	D-20	D-21
Deliveries second tier	4	1	1	0	0	1	–
Deliveries third tier	1	1	0	0	1		–
First tier component inventory	7	4	1	1	1	2	–
Second tier component inventory	2	2	1	1	2	1	–

GP2 Results, Case B, Stranger

Day	D-1	D-2	D-3	D-4	D-44	D-6	D-7
Deliveries second tier	14	0	7	14	0	14	0
Deliveries third tier	10	10	0	10	10	0	20
First tier component inventory	15	5	2	12	4	11	2
Second tier component inventory	6	16	9	5	15	1	21
Day	D-8	D-9	D-10	D-11	D-12	D-13	D-14
Deliveries second tier	14	0	14	0	0	7	7
Deliveries third tier	0	10	0	10	0	10	0
First tier component inventory	11	1	15	12	9	10	9
Second tier component inventory	7	17	3	13	13	16	9

GP2 Results, Case B, Stranger, continued

Day	D-15	D-16	D-17	D-18	D-19	D-20	D-21
Deliveries second tier	7	0	7	7	0	0	–
Deliveries third tier	0	10	0	10	0	0	–
First tier component inventory	7	3	6	13	13	13	–
Second tier component inventory	2	12	5	8	8	8	–

Annex A: Output from the GP Trials

GP3 Results, Case B, Stranger

Day	D-1	D-2	D-3	D-4	D-44	D-6	D-7
Deliveries second tier	14	0	7	14	0	14	0
Deliveries third tier	10	10	0	10	10	0	20
First tier component inventory	15	5	2	12	4	11	2
Second tier component inventory	6	16	9	5	15	1	21
Day	D-8	D-9	D-10	D-11	D-12	D-13	D-14
Deliveries second tier	14	0	14	0	0	7	7
Deliveries third tier	0	10	0	10	0	10	0
First tier component inventory	11	1	15	12	9	10	9
Second tier component inventory	7	17	3	13	13	16	9
Day	D-15	D-16	D-17	D-18	D-19	D-20	D-21
Deliveries second tier	7	7	0	7	0	0	–
Deliveries third tier	0	10	10	0	0	0	–
First tier component inventory	7	10	6	13	13	13	–
Second tier component inventory	2	5	15	8	8	8	–

Case C—High volume supply chain. Runner

GP composite file requirements list = Final call-off file requirements file

Day	D-1	D-2	D-3	D-4	D-44	D-6	D-7
Requirements	350	304	292	316	274	0	260
Day	D-8	D-9	D-10	D-11	D-12	D-13	D-14
Requirements	306	274	314	0	212	190	194
Day	D-15	D-16	D-17	D-18	D-19	D-20	D-21
Requirements	210	246	0	200	200	196	196
Day	D-22	D-23	D-24	D-25	D-26	D-27	D-28
Requirements	244	0	210	186	208	228	252
Day	D-29	D-30	D-31	D-32	D-33	D-34	D-35
Requirements	0	260	192	200	192	196	200
Day	D-36	D-37	D-38	D-39	D-40	D-41	D-42
Requirements	184	198	202	212	244	250	248
Day	D-43	D-44					
Requirements	284	212					

GP1 Results, Case C, Runner

Day	D-1	D-2	D-3	D-4	D-44	D-6	D-7
Deliveries second tier	304	292	316	274	0	260	306
First tier component inventory	305	293	317	275	1	261	307
Second tier component inventory	339	363	321	47	307	353	321
Day	D-8	D-9	D-10	D-11	D-12	D-13	D-14
Deliveries second tier	274	314	0	212	190	194	210
First tier component inventory	275	315	1	213	191	195	211
Second tier component inventory	361	47	259	237	241	257	293
Day	D-15	D-16	D-17	D-18	D-19	D-20	D-21
Deliveries second tier	246	0	200	200	196	196	244
First tier component inventory	247	1	201	201	197	197	245
Second tier component inventory	47	247	247	243	243	291	47

GP1 Results, Case C, Runner, continued

Day	D-22	D-23	D-24	D-25	D-26	D-27	D-28
Deliveries second tier	0	210	186	208	228	252	0
First tier component inventory	1	7	187	209	229	253	1
Second tier component inventory	257	233	255	275	299	47	307
Day	D-29	D-30	D-31	D-32	D-33	D-34	D-35
Deliveries second tier	260	192	200	192	196	200	184
First tier component inventory	261	193	201	193	197	201	185
Second tier component inventory	239	247	239	243	247	231	245
Day	D-36	D-37	D-38	D-39	D-40	D-41	D-42
Deliveries second tier	198	202	212	244	250	248	284
First tier component inventory	199	203	213	245	251	249	285
Second tier component inventory	249	259	291	297	295	331	259
Day	D-43						
Deliveries second tier	212						
First tier component inventory	213						
Second tier component inventory	47						

GP2 Results, Case C, Runner

Day	D-1	D-2	D-3	D-4	D-44	D-6	D-7
Deliveries second tier	180	360	324	252	36	288	288
First tier component inventory	230	286	318	254	16	304	332
Second tier component inventory	375	215	241	39	503	315	277

(continued)

Annex A: Output from the GP Trials

(continued)

Day	D-8	D-9	D-10	D-11	D-12	D-13	D-14
Deliveries second tier	216	360	0	468	216	0	180
First tier component inventory	242	328	14	482	486	296	282
Second tier component inventory	361	1	501	33	267	467	337
Day	D-15	D-16	D-17	D-18	D-19	D-20	D-21
Deliveries second tier	252	0	396	144	72	180	252
First tier component inventory	324	78	474	418	290	274	330
Second tier component inventory	85	485	89	345	423	343	91

GP2 Results, Case C, Runner, continued

Day	D-22	D-23	D-24	D-25	D-26	D-27	D-28
Deliveries second tier	0	396	180	36	216	216	0
First tier component inventory	86	482	452	302	310	298	46
Second tier component inventory	491	95	315	479	313	97	547
Day	D-29	D-30	D-31	D-32	D-33	D-34	D-35
Deliveries second tier	504	216	0	144	216	216	252
First tier component inventory	550	506	314	258	282	302	354
Second tier component inventory	43	277	477	333	267	251	249
Day	D-36	D-37	D-38	D-39	D-40	D-41	D-42
Deliveries second tier	108	216	180	252	180	288	288
First tier component inventory	278	296	274	314	250	288	328
Second tier component inventory	391	275	345	293	363	275	187
Day	D-43						
Deliveries second tier	180						
First tier component inventory	224						
Second tier component inventory	107						

GP3 Results, Case C, Runner

Day	D-1	D-2	D-3	D-4	D-44	D-6	D-7
Deliveries second tier	180	360	324	252	36	540	36
First tier component inventory	230	286	318	254	16	556	332
Second tier component inventory	375	215	241	39	503	63	527
Day	D-8	D-9	D-10	D-11	D-12	D-13	D-14
Deliveries second tier	216	360	0	468	0	216	180
First tier component inventory	242	328	14	482	270	296	282
Second tier component inventory	361	1	501	33	483	267	337
Day	D-15	D-16	D-17	D-18	D-19	D-20	D-21
Deliveries second tier	252	0	396	0	216	180	252
First tier component inventory	324	78	474	274	290	274	330
Second tier component inventory	85	485	89	489	273	343	91

GP3 Results, Case C, Runner, continued

Day	D-22	D-23	D-24	D-25	D-26	D-27	D-28
Deliveries second tier	0	396	0	216	216	216	0
First tier component inventory	86	482	272	302	310	298	46
Second tier component inventory	491	95	495	279	313	97	547
Day	D-29	D-30	D-31	D-32	D-33	D-34	D-35
Deliveries second tier	504	0	180	180	216	216	180
First tier component inventory	550	290	278	258	282	302	282
Second tier component inventory	43	493	313	283	267	251	321
Day	D-36	D-37	D-38	D-39	D-40	D-41	D-42
Deliveries second tier	180	216	180	252	216	252	288
First tier component inventory	278	296	274	314	286	288	328
Second tier component inventory	291	275	345	293	327	325	187
Day	D-43						
Deliveries second tier	180						
First tier component inventory	224						
Second tier component inventory	107						

Case C—High volume supply chain. Stranger

GP composite file requirements list = Final call-off file requirements file

Day	D-1	D-2	D-3	D-4	D-44	D-6	D-7
Requirements	58	46	70	66	54	0	56
Day	D-8	D-9	D-10	D-11	D-12	D-13	D-14
Requirements	54	50	68	0	46	54	40
Day	D-15	D-16	D-17	D-18	D-19	D-20	D-21
Requirements	60	46	0	38	30	38	38
Day	D-22	D-23	D-24	D-25	D-26	D-27	D-28
Requirements	42	0	42	36	36	34	48
Day	D-29	D-30	D-31	D-32	D-33	D-34	D-35
Requirements	0	38	56	36	54	34	50

GP composite file requirements list = Final call-off file requirements file (cont.)

Day	D-36	D-37	D-38	D-39	D-40	D-41	D-42
Requirements	48	54	46	52	46	54	42
Day	D-43	D-44					
Requirements	50	36					

Annex A: Output from the GP Trials

GP1 Results, Case C, Stranger

Day	D-1	D-2	D-3	D-4	D-44	D-6	D-7
Deliveries second tier	46	70	66	54	0	56	54
First tier component inventory	47	71	67	55	1	57	55
Second tier component inventory	74	70	58	4	60	58	54
Day	D-8	D-9	D-10	D-11	D-12	D-13	D-14
Deliveries second tier	50	68	0	46	54	40	60
First tier component inventory	51	69	1	47	55	41	61
Second tier component inventory	72	4	50	58	44	64	50
Day	D-15	D-16	D-17	D-18	D-19	D-20	D-21
Deliveries second tier	46	0	38	30	38	38	42
First tier component inventory	47	1	39	31	39	39	43
Second tier component inventory	4	42	34	42	42	46	4
Day	D-22	D-23	D-24	D-25	D-26	D-27	D-28
Deliveries second tier	0	42	36	36	34	48	0
First tier component inventory	1	43	37	37	35	49	1
Second tier component inventory	46	40	40	38	52	4	42
Day	D-29	D-30	D-31	D-32	D-33	D-34	D-35
Deliveries second tier	38	56	36	54	34	50	48
First tier component inventory	39	57	37	55	35	51	49
Second tier component inventory	60	40	58	38	54	52	58
Day	D-36	D-37	D-38	D-39	D-40	D-41	D-42
Deliveries second tier	54	46	52	46	54	42	50
First tier component inventory	55	47	53	47	55	43	51
Second tier component inventory	50	56	50	58	46	54	40

GP1 Results, case C, Stranger, continued

Day	D-43
Deliveries second tier	36
First tier component inventory	37
Second tier component inventory	4

GP2 Results, Case C, Stranger

Day	D-1	D-2	D-3	D-4	D-44	D-6	D-7
Deliveries second tier	36	72	72	36	36	36	72
First tier component inventory	33	59	61	31	13	49	65
Second tier component inventory	114	92	70	34	98	112	40

(continued)

(continued)

Day	D-8	D-9	D-10	D-11	D-12	D-13	D-14
Deliveries second tier	36	72	0	108	36	0	72
First tier component inventory	47	69	1	109	99	45	77
Second tier component inventory	104	32	132	24	88	138	66
Day	D-15	D-16	D-17	D-18	D-19	D-20	D-21
Deliveries second tier	36	0	72	0	36	36	36
First tier component inventory	53	7	79	41	47	45	43
Second tier component inventory	30	130	58	108	72	86	50
Day	D-22	D-23	D-24	D-25	D-26	D-27	D-28
Deliveries second tier	0	108	0	36	0	72	0
First tier component inventory	1	109	67	67	31	69	21
Second tier component inventory	150	42	142	106	156	84	134
Day	D-29	D-30	D-31	D-32	D-33	D-34	D-35
Deliveries second tier	72	36	0	72	0	72	36
First tier component inventory	93	91	35	71	17	55	41
Second tier component inventory	62	76	126	54	154	82	96
Day	D-36	D-37	D-38	D-39	D-40	D-41	D-42
Deliveries second tier	72	36	72	36	0	72	72
First tier component inventory	65	47	73	57	11	29	59
Second tier component inventory	74	88	66	80	130	58	36

GP2 Results, Case C, Stranger, continued

Day	D-43
Deliveries second tier	36
First tier component inventory	45
Second tier component inventory	50

GP3 Results, Case C, Stranger

Day	D-1	D-2	D-3	D-4	D-44	D-6	D-7
Deliveries second tier	36	72	72	36	36	108	0
First tier component inventory	33	59	61	31	13	121	65
Second tier component inventory	114	92	70	34	98	40	140
Day	D-8	D-9	D-10	D-11	D-12	D-13	D-14
Deliveries second tier	36	72	0	108	0	36	72
First tier component inventory	47	69	1	109	63	45	77
Second tier component inventory	104	32	132	24	124	88	66
Day	D-15	D-16	D-17	D-18	D-19	D-20	D-21
Deliveries second tier	36	0	72	0	36	36	36

(continued)

Annex A: Output from the GP Trials

(continued)

First tier component inventory	53	7	79	41	47	45	43
Second tier component inventory	30	130	58	108	72	86	50
Day	D-22	D-23	D-24	D-25	D-26	D-27	D-28
Deliveries second tier	0	108	0	36	0	72	0
First tier component inventory	1	109	67	67	31	69	21
Second tier component inventory	150	42	142	106	156	84	134
Day	D-29	D-30	D-31	D-32	D-33	D-34	D-35
Deliveries second tier	72	0	36	72	0	72	36
First tier component inventory	93	55	35	71	17	55	41
Second tier component inventory	62	112	126	54	154	82	96
Day	D-36	D-37	D-38	D-39	D-40	D-41	D-42
Deliveries second tier	72	36	72	36	36	36	72
First tier component inventory	65	47	73	57	47	29	59
Second tier component inventory	74	88	66	80	94	108	36

GP3 Results, Case C, Stranger, continued

Day	D-43
Deliveries second tier	36
First tier component inventory	45
Second tier component inventory	50

Annex B—Extending Synchronisation in the Supply Chain

Case C. Composite file requirements list = Final call-off file requirements list

GP composite file requirements list = Final call-off file requirements file

Day	D-1	D-2	D-3	D-4	D-44	D-6	D-7
Requirements	58	46	70	66	54	0	56
Day	D-8	D-9	D-10	D-11	D-12	D-13	D-14
Requirements	54	50	68	0	46	54	40
Day	D-15	D-16	D-17	D-18	D-19	D-20	D-21
Requirements	60	46	0	38	30	38	38
Day	D-22	D-23	D-24	D-25	D-26	D-27	D-28
Requirements	42	0	42	36	36	34	48
Day	D-29	D-30	D-31	D-32	D-33	D-34	D-35
Requirements	0	38	56	36	54	34	50

(continued)

(continued)

Day	D-36	D-37	D-38	D-39	D-40	D-41	D-42
Requirements	48	54	46	52	46	54	42
Day	D-43	D-44					
Requirements	50	36					

Non-sequenced, second-tier deliveries

Day	D-1	D-2	D-3	D-4	D-44	D-6	D-7
Deliveries second tier	0	36	72	72	36	36	108
Deliveries third tier	100	100	50	50	0	100	50
First tier component inventory	33	59	61	31	13	121	65
Second tier component inventory	106	84	62	26	90	32	132
Day	D-8	D-9	D-10	D-11	D-12	D-13	D-14
Deliveries second tier	0	36	72	0	108	0	36
Deliveries third tier	100	0	0	100	0	100	50
First tier component inventory	47	69	1	109	63	45	77
Second tier component inventory	96	24	124	16	116	130	58

Non-sequenced, second-tier deliveries, continued

Day	D-15	D-16	D-17	D-18	D-19	D-20	D-21
Deliveries second tier	72	36	0	72	0	36	36
Deliveries third tier	0	50	50	0	50	0	50
First tier component inventory	53	7	79	41	47	45	43
Second tier component inventory	72	122	50	100	64	78	42
Day	D-22	D-23	D-24	D-25	D-26	D-27	D-28
Deliveries second tier	36	0	108	0	36	0	72
Deliveries third tier	0	100	0	100	0	50	0
First tier component inventory	1	109	67	67	31	69	21
Second tier component inventory	142	34	134	98	148	76	126
Day	D-29	D-30	D-31	D-32	D-33	D-34	D-35
Deliveries second tier	0	72	0	36	72	0	72
Deliveries third tier	50	0	50	50	0	100	0
First tier component inventory	93	55	35	71	17	55	41
Second tier component inventory	54	104	118	46	146	74	88
Day	D-36	D-37	D-38	D-39	D-40	D-41	D-42
Deliveries second tier	36	72	36	72	36	36	36
Deliveries third tier	50	50	50	50	100	0	50
First tier component inventory	65	47	73	57	47	29	59
Second tier component inventory	66	80	58	122	86	100	28

(continued)

Annex B—Extending Synchronisation in the Supply Chain 191

(continued)

Day	D-43	D-44
Deliveries second tier	72	36
Deliveries third tier	0	50
First tier component inventory	45	
Second tier component inventory	42	

Sequenced, second-tier deliveries

Day	D-1	D-2	D-3	D-4	D-44	D-6	D-7
Deliveries second tier	58	46	70	66	54	0	56
Deliveries third tier	50	100	0	100	0	100	0
First tier component inventory	1	1	1	1	1	1	1
Second tier component inventory	76	130	60	94	40	140	84
Day	D-8	D-9	D-10	D-11	D-12	D-13	D-14
Deliveries second tier	54	50	68	0	46	54	40
Deliveries third tier	50	50	0	100	50	0	100
First tier component inventory	1	1	1	1	1	1	1
Second tier component inventory	80	80	12	112	116	62	122
Day	D-15	D-16	D-17	D-18	D-19	D-20	D-21
Deliveries second tier	60	46	0	38	30	38	38
Deliveries third tier	0	0	100	0	50	0	50
First tier component inventory	1	1	1	1	1	1	1
Second tier component inventory	62	16	116	78	98	60	72
Day	D-22	D-23	D-24	D-25	D-26	D-27	D-28
Deliveries second tier	42	0	42	36	36	34	48
Deliveries third tier	0	100	0	50	0	50	0
First tier component inventory	1	1	1	1	1	1	1
Second tier component inventory	30	130	88	102	66	82	34
Day	D-29	D-30	D-31	D-32	D-33	D-34	D-35
Deliveries second tier	0	38	56	36	54	34	50
Deliveries third tier	100	0	0	100	0	100	0
First tier component inventory	1	1	1	1	1	1	1
Second tier component inventory	134	96	40	104	50	116	66
Day	D-36	D-37	D-38	D-39	D-40	D-41	D-42
Deliveries second tier	48	54	46	52	46	54	42
Deliveries third tier	50	50	50	50	50	50	50
First tier component inventory	1	1	1	1	1	1	1
Second tier component inventory	68	64	68	66	70	66	74

Sequenced, second-tier deliveries, continued

Day	D-43	D-44
Deliveries second tier	50	36
Deliveries third tier	50	0
First tier component inventory	1	1
Second tier component inventory	74	36

References

Coleman J, Lyons A, Kehoe D (2004) The glass pipeline: increasing supply chain synchronisation through information transparency. Int J Technol Manag 28(2):172–190

Holweg M, Bicheno J (2002) Supply chain simulation: a tool for education, enhancement and endeavour. Int J Prod Econ 78:163–175

Lee HL, Padmanabhan V, Whang S (1997) Information distortion in a supply chain: The bullwhip effect. Manag Sci 43(4):546–558

Lyons A, Coronado Mondragon A, Bremang A, Kehoe D, Coleman J (2005) Prototyping an information system's requirements architecture for customer-driven, supply-chain operations. Int J Prod Res 43(20):4289–4319

Mason-Jones R, Towill DR (1999) Using the information de-coupling point to improve supply chain performance. Int J Logist Manag 10(2):13–26

Mondragon AEC, Lyons AC (2008) Investigating the implications of extending synchronized sequencing in automotive supply chains: The case of suppliers in the European automotive sector. Int J Prod Res 46(11):2867–2888

Olhager J (2002) Supply chain management: a just-in-time perspective. Prod Plan Control 13(8):681–687

Persson F, Olhager J (2002) Performance simulation of supply chain designs. Int J Prod Econ 77:231–245

Index

121Time.com, 89

A
Adidas, 87–89
Agility, 15, 46, 51–53, 56, 67, 68, 70
Alignment of production with demand, 22, 23, 35, 36, 41–44, 48, 154, 156, 157
ANSI, 58
ARIS, 20, 24, 108, 119, 120, 131

B
Backorders, 126, 135, 142, 143, 146, 147, 153, 159, 162, 175
Behavioural measures, 135, 137, 139, 147
BMW, 72, 84, 107
Build-to-order, 44, 61, 69, 88, 161, 173, 174
Bullwhip effect, 49, 69, 70, 101, 135–137, 148, 158, 166, 192
Bullwhip index, 135, 137, 173, 174
Burns, 53–55, 70, 84, 94
Business process management, 118, 130, 131
Business process modelling language (BPML), 123
Business processes, 21, 62, 95, 114–118, 120–123

C
Changeover, 6, 9, 23, 24, 26, 33, 36, 37, 41, 44, 51, 154, 168
Chesbrough, 106, 109, 110
Choice navigation, 78, 85, 86, 88, 89, 91
CIMOSA, 119, 120

Co-design, 79, 85–89, 91, 93, 94
Configuration toolkit, 85, 92
Cost measures, 135, 143, 144, 146
Creative involvement of the workforce, 22, 23, 42–44, 48, 154, 155
Customer-driven guidelines, 23, 153, 155–157, 174
Customer-driven practices, 23, 155, 167, 175
Customer-driven processes, 22, 43, 45, 48, 56, 149, 153–155, 167, 169, 173
Customer-led innovation, 107
Customer-order decoupling point, 51–53, 68

D
Data Flow Diagram (DFD), 121
Demand amplification, 46–48, 68, 136, 137
Demand forecast MAD, 135
DoDAF, 119, 120

E
EDI, 58–62, 67, 69, 130, 151, 157, 159
EDIFACT, 58
Efficient supply chain, 13, 14, 16
Elimination of waste, 16, 22, 23, 42, 43, 154, 156
Entity/Relationship Diagram (ERD), 121
ERP, 59–62, 65, 67
External networking, 107
Energic strategy, 10, 13

F
Facility layout, 24, 35
Fisher, 13–15, 17, 19, 56

F (cont.)
Fit, 79
Form, 79
Functional product, 13–16, 18
Functionality, 16, 61, 71, 79, 80, 87

G
General Packet Radio Services, 64
Generic strategy, 10, 18
Geographic Information Systems, 64
GERAM, 119, 120, 130
GIM, 119, 120
Glass pipeline, 49
Global Positioning Systems, 64

H
Hill, 10–13, 19, 68, 70
Horizontal collaboration, 102–105, 109, 110, 149
Huang, 15, 16, 19

I
IDEF, 119, 121, 122, 131
Innovative product, 13–16, 18, 108, 109
Integration of suppliers, 22, 23, 41–43, 48, 154, 156
Intellectual property, 8, 106
Inventory holding cost, 84, 135, 144, 147
Inventory level, 59, 65, 100, 140, 141, 153, 163, 165, 167, 171, 172

K
Kaizen, 23, 27, 42–44, 90, 118, 154
Kanban, 27, 38, 39, 41, 171, 172

L
Ladder diagram, 153, 155–157, 167–170
Leagile, 19, 56
Lean, 15, 17

M
Make-or-buy decision, 7, 9, 12, 13, 18, 19, 115
Mass customisation, 10, 11, 16, 52–56, 71–74, 76–92, 94
Michaelides, 65, 66, 69
Modularity, 54, 57, 58, 82, 84, 92, 93, 97
MRP, 59, 60, 61, 157, 160, 165
Multi-agent systems, 128

O
Object-Oriented Methods (OOM), 123
Open innovation, 106–108, 110, 111
Outsourcing, 8, 53, 55, 68, 96, 97, 99–101, 150
Overall Equipment Effectiveness (OEE), 34

P
Pandora radio, 72, 77, 82
Paradox of choice, 94
PERA, 119, 120, 131
Postponement, 52–55, 57, 58, 69, 70, 84, 94
Process flow chart, 26
Process flow diagram, 25
Procter & Gamble, 47, 90
Product life cycle, 16–18
Pull, 23, 27, 37, 38, 41, 44, 51, 53, 58
Push, 27, 37, 57, 58

R
Relative demand volatility, 56
Reliability measures, 135, 142, 143, 146, 147, 153, 159, 162
Repeater, 6, 7, 13
Responsive supply chain, 13, 14, 16, 100
Responsiveness measures, 135, 139, 141, 144, 145, 153, 159, 160, 162, 164, 166, 167
RFID, 64, 67, 68
Robust process design, 78, 83, 85
Rope analogy, 37
Rother, 27, 44
Runner, 6, 7, 13, 152, 153, 158, 159, 161, 162, 164–167, 175–177, 179–181, 183–186

S
Sears, 73, 77, 88
Selve, 72
Sequenced supply, 66, 67, 109, 162, 171, 172
Shirose, 34, 44
Shook, 27, 44
Solution space development, 77, 78
Spreadsheet simulation, 125
Statistical process control, 23, 32
Stockouts, 11, 13–15, 135, 143, 147, 174
Stranger, 6, 7, 13, 152, 153, 158, 159, 161, 162, 164–167, 173, 177, 178, 179, 181–183, 186–189
Supplier park, 66, 96–101, 109, 149, 150
Supply chain cycle time, 135, 139–142, 145, 147, 153, 158, 159, 162, 164, 166, 167
Supply chain modelling, 121, 123

Supply chain performance
  measurement, 133–136, 138, 140, 142, 144, 146–148
Supply chain simulation, 124, 125, 127, 129, 130, 192
Supply chain strategy, 2–6, 8, 10, 12–14, 16, 18
Supply chain structures, 4, 95, 114, 115, 124
Supply chain topologies, 114
Synchronisation index, 135, 138–140, 145, 153, 158, 159, 162, 164–167, 174
Synchronisation, 21, 22, 43, 45–49, 51–53, 63, 66, 67, 135, 138–140, 145, 147–150, 153, 158, 159, 162–167, 169–171, 173–175, 189, 191, 192
System dynamics, 125

T
Takt time, 39–41, 44, 145–147, 154
TOGAF, 119, 120
Total productive maintenance, 23, 33
Transportation cost, 97, 134, 135, 144

U
Unified Modelling Language (UML), 123

V
Value-adding contribution, 135, 139, 142, 147
Value stream map, 23, 25–27, 43, 44, 145–147
Vendor-managed inventory (VMI), 50
Venturing, 108–110
Vertical collaboration, 102

W
WEBEDI, 62
Workplace organisation, 23, 34, 43, 154

Y
Yamazumi, 39, 40
Yang, 55, 70, 84, 94

Z
Zachman, 119, 120, 131
Zafu, 88, 89
Zara, 56, 100–102, 109